Helion & Company Limited
Unit 8 Amherst Business Centre
Budbrooke Road
Warwick
CV34 5WE
England
Tel. 01926 499 619
Email: info@helion.co.uk
Website: www.helion.co.uk
Twitter: @helionbooks
Visit our blog http://blog.helion.co.uk/

Designed & typeset by Farr out Publications, Wokingham, Berkshire
Cover design Paul Hewitt, Battlefield Design (www.battlefield-design.co.uk)

ISBN 978-1-913336-31-8

British Library Cataloguing-in-Publication Data
A catalogue record for this book is available from the British Library

We always welcome receiving book proposals from prospective authors.

CONTENTS

Note: In order to simplify the use of this book, all names, locations and geographic designations are as provided in *The Times World Atlas*, or other traditionally accepted major sources of reference, as of the time of described events.

ABBREVIATIONS

ASCC Air Standardisation Coordinating Committee
AV-MF Aviatsiya Voyenno-Morskogo Flota (Soviet Naval Aviation)
CIA Central Intelligence Agency
C-in-C Commander in chief
c/n Construction number
CO Commanding officer
DA Dalnyaya Aviatsiya (Soviet Long-Range Aviation)
HQ Headquarters
ICBM Intercontinental Ballistic Missile
Il Ilyushin (the design bureau headed by Sergey Ilyushin)
kt Kiloton – a unit of energy equal to 4.184 terajoules (4.184×10^{12} J), conventionally given as an equivalent of the energy released by a chemical explosion of one thousand tonnes TNT
La Lavochkin (the design bureau headed by Semyon Lavochkin), also known as NPO Lavochkin or OKB-301
LTU Letno-Takticheskiye Ucheniya (Flying Tactical Exercises)
M Myasishchev (the design bureau led by Vladimir Myasishchev, also known as OKB-23)
MRAP Morskoy Raketniy Aviatsionniy Polk (Naval Rocket (missile) Aviation Regiment)
MiG Mikoyan i Gurevich (the design bureau led by Artyom Ivanovich Mikoyan and Mikhail Iosifovich Gurevich, also known as OKB-155 or MMZ 'Zenit')
Mt Megaton – is a unit of energy equal to 4.184 petajoules (4.184×10^{15} J), conventionally given as an equivalent of the energy released by a chemical explosion of one million tonnes TNT
NATO North Atlantic Treaty Organization
OAE Otdelnaya Aviatsionnaya Eskadrilya (Independent Aviation Squadron)
ODRAP Otdelnyy Dalnerazvedyvatelnyy Aviatsionnyy Polk (Independent Long-Range Reconnaissance Aviation Regiment)
OTBAK Otdelnyy Tyazhelyy Bombardirovochnyy Aviatsionniy Korpus (Independent Heavy Bomber Aviation Corps)
TBAD Tyazholaya Bombardirovochnaya Aviatsionnaya Diviziya (Heavy Bomber Aviation Division)
TBAP Tyazhelyy Bombardirovochnyy Aviatsionnyy Polk (Heavy Bomber Aviation Regiment)
Tu Tupolev (the design bureau headed by Andrei Tupolev, also known as OKB-156)
US/USA United States of America
USAF United States Air Force
USSR Union of Soviet Socialist Republics
Yak Yakovlev (the design bureau led by Alexander Yakovlev, also known as OKB-115 or MMZ 'Skorost')

INTRODUCTION

The development as well as testing of nuclear weapons in the Union of Soviet Socialist Republics (USSR, colloquially 'Soviet Union') are by and large known and have been described in detail on numerous occasions. Yet, 'in detail' does not necessarily mean 'exhaustingly', for when getting acquainted with most English language sources regarding this subject it is impossible not to notice that certain aspects of Soviet nuclear weapons development are treated in a rather superficial manner. It has to be pointed out that producing a workable nuclear device and detonating it constitute major achievements by any standard, at the same time however for this to be of practical military value a method of deliverly has to be tested. In the case of nuclear weapons, in particular when ballistic missiles were still a novelty with limited capabilities, the best option for their delivery was that of an aerial bomb which was to be dropped from an aircraft. It was no different with the Soviets, who built nuclear bombs and tested them by dropping them from bombers. Yet, even though this was crucial, this aspect of Soviet nuclear weapons development is given little consideration. In the vast majority of cases anyone interested must be satisfied with a few brief lines stating that a nuclear device was tested by being air dropped on a given date, with the aircraft type sometimes denoted. Similarly, when dealing with Soviet aircraft – bombers in particular – numerous monographs are available which sometimes go into great technical detail when describing a given aircraft type and its various sub-types, yet more often than not shy away from anything but a brief mention of how it was utilised in the testing of nuclear weapons. The present volume attempts to make up for the shortcomings outlined above. It neither strives to give a detailed description of Soviet nuclear weapons development nor the development history of Soviet bomber types but instead aims to give a succinct account of how these 'doomsday devices' were practically tested by means of an airdrop. Short of an actual nuclear war, which we were mercifully spared, this is also the closest to providing a picture of how nuclear warfare might have been conducted. It is not an uncommonly held belief that introductions are good when they are brief, for the reader should not be unnecessarily delayed from accessing the essence. The author shares this opinion and therefore without further ado presents the history of the first Soviet nuclear bomb airdrop as well as those which subsequently followed.

1
DROPPING THE BIG ONE

In some circles 'the Big One' colloquially refers to a nuclear bomb. However, before a nuclear bomb could be airdropped the Soviets had to have one to begin with. Much has been written and published about Soviet efforts to build a nuclear weapon and it seems there is little one can add. Yet – even without going much into detail – it still seems indispensable to devote at least a few words to how the red 'doomsday device' came into being before its testing by means of an airdrop can be recounted.

Joe's Bomb

Soviet research on the subject of nuclear physics started in the 1920s, virtually as soon as the state of the country which has been ravaged by a world war, two revolutions and a civil war, allowed for it. By 1940 the first proposals for a nuclear weapon were put forward, but at that time the idea was not fully appreciated by top decision makers and the necessary infrastructure to build a nuclear bomb did not yet physically exist.[1] In 1940-1942, Russian physicist Georgy Flyorov, then serving as an officer of the Soviet Air Force, noted that the leading German, British and then American scientists had all ceased publishing papers on nuclear science. This made it clear that all three countries had an active – yet secret – nuclear research program.[2] The German invasion in June 1941 obviously did not help to advance the issue. However, in September 1942 the matter was taken up by the State Defence Committee which ordered scientists to 'organize work with uranium'.[3] This can arguably be seen as the start in earnest of the Soviet nuclear weapons project, yet for a number of reasons the resulting efforts failed to deliver tangible results before the end of the war.

The US atomic bombing of Hiroshima and Nagasaki, which demonstrated the previously almost unimaginable power of nuclear weapons, was crucial in reinvigorating the Soviet nuclear weapons program. It became obvious that a nuclear weapon had to be built – no matter the cost – and any and all means were allowed to acquire everything (know-how included) to accomplish that task. Much was obtained in occupied Germany which before its capitulation had made considerable strides in nuclear research.[4] Significant amounts of scientific instruments, technical equipment, machinery and documents were seized by the Soviets, while a number of German nuclear scientists were made offers of the kind they could not refuse. As has been confirmed by the declassified Venona files even more vital information was gained through British and American spies (not a few were in fact willing informers), who worked on the US nuclear bomb development.[5]

Thus, after the end of the Second World War the Soviets were able to intensively pursue a nuclear weapons program. Vital data about German nuclear research and especially about prior American nuclear bomb development obtained through intelligence allowed the Soviets to leapfrog through stages of development, to avoid dead ends and to focus on solutions which were known to work in practice. As a result, the Soviet effort headed by Igor Kurchatov produced the desired outcome even before the end of the 1940s, with the first Soviet nuclear device exploded on 29 August 1949 at the Semipalatinsk Test Site in Kazakhstan.[6] It was an implosion lens design utilising plutonium which produced a yield of 22kt when detonated. For the purpose of the test the said device was placed on a 30m high tower with the coordinates of the site of the first Soviet nuclear explosion being given as 50.43794°N/77.81409°E. The Soviet 'doomsday machine' was designated RDS-1 with various explanations for the acronym given (some appearing as late as the 1990s and thus likely being post-factum inventions) while in the West it became known as Joe-1 in a not particularly veiled reference to the Soviet dictator Joseph Stalin.

A Suitable Test Site

When a nuclear detonation takes place, it creates a giant fireball. The latter is extremely hot and therefore swiftly rises high into the air becoming the cap of a mushroom cloud. While rising, the fireball creates a powerful updraft which picks up dust and smoke, in turn forming the mushroom cloud's stem. Thus, as a result of nuclear

A Soviet test rig. One similar to this would have held the device that was detonated in the test codenamed *Pervaya Molniya* (First Lightning) on 29 August 1949. The test-explosion yield was the equivalent of 22kt of TNT. Upon recovering some of the debris with help of a Boeing WB-29 weather reconnaissance aircraft on 1 September 1949, the surprised US intelligence community concluded that the USSR was now a nuclear power and codenamed the bomb 'Joe-1'.

The mushroom cloud caused by the 'Joe-1' test in 1949. (Still from a Soviet documentary film)

RDS-1: the first 'operational' Soviet nuclear bomb, on display at the *Muzey yadernogo oruzhiya* (Nuclear Weapons Museum) in Sarov. Obviously, its design was very similar to that of the US 'Fat Man' plutonium bomb of 1945.

explosions gigantic mushrooms grew over the landscape, this only coming to an end when atmospheric nuclear testing was banned. Even then, underground nuclear explosion still tormented the earth. Some places were unfortunate enough to experience this on an almost unimaginable scale.

Semipalatinsk

In 1947 the *2-y Gosudarstvennyy tsentralnyy nauchno-issledovatelskiy ispytatelny* polygon, that is the 2nd State Central Research Testing Ground, was established in the Kazakh Soviet Republic (which nowadays is an independent state). Also going by other designations and colloquial names it is most often referred to as the *Semipalaatinskiy ispytatelny* polygon, that is the Semipalatinsk Test Site (STS). It was located about 130km north-west of Semipalatinsk City on the left bank of the Irtysh river. The location was chosen because of its remoteness, which facilitated secrecy, as well as for being sparsely populated. The STS had an area of 18,500 square kilometres. On its grounds there was the so called Opytnoye pole – that is a trail or test field – which was in turn divided into *ploshchadki* – that is sites where the actual nuclear tests were conducted. The tests required the construction of various facilities and installations, and in addition entire towns were built to house the military and civilian personnel operating the STS and their families. In all 456 nuclear and thermonuclear explosions took place at the STS, comprising 116 atmospheric detonations (tower, air drop and missile) as well 340 underground ones (borehole, shaft and tunnel). The final nuclear test there took place on 19 October 1989 and on 29 August 1991 the site was formally closed.

However, STS's closure was not an end to the matter, and certain effects lingered on and persist to this very day. Moreover, the end of the testing was at the same time the beginning of efforts to deal with its aftermath. Much could be written about the adverse effects on health that nuclear testing at the STS had, but it is sufficient to say that radioactive fallout had a direct impact (including cancerogenic effects) on the health of about 200,000 people and due to genetic defects STS's legacy is here to stay literally for generations to come.

There were other problems too. Namely the Soviet Union was dissolved by the Belovezha Accords signed on 8 December 1991 but in many places the formal dissolution was both preceeded and followed by a de facto collapse. This collapse affected STS where fissile material was simply abandoned in tunnels, bore holes and the like used for testing and it was a realistic possibility that it might fall into the hands of criminals and terrorists. However, the period when all this dangerous material was left virtually unguarded was relatively short. Kazakhstan, Russia and the United Sates soon launched a joint effort to secure the radioactive material left behind. The operation was completed by October 2012 and it appears that no unauthorised persons were able to lay hands upon it.

At the time of writing, the former STS is administered by the National Nuclear Centre of Kazakhstan. Since 2014 large parts of what used to be the testing grounds have been opened to economic activity such as mining of natural resources, but also perhaps a little surprisingly, tourism and even agriculture.

Novaya Zemlya

Despite their vastness, the expanses of Kazakhstan did not provide enough space for live trials of nuclear and thermonuclear weapons with a large yield. Furthermore, the location was obviously unsuitable for the testing of naval nuclear weapons. For this reason, Novaya Zemlya, an archipelago in the Arctic waters of northern Russia at the extreme northeast of Europe was selected as an additional location for the conduct of nuclear trials. The test site was formally opened on 17 September 1954 and was at that time known under an ambiguous coded designation *Obyekt-700* (Object-700). Large areas or zones were taken over for the testing: Zone A (Chyornaya Guba Bay); Zone B (Matochkin Shar Strait); and Zone C (Sukhoy Nos Cape). The fourth, Zone D (Mityushikha Guba Bay), was established in 1956. Overall, the test site's land area was 55,200 square kilometres with an additional 36,000 square kilometres of adjoining waters. Between the years 1955 and 1990 at least 132 (some sources put the number at well over 200) nuclear and thermonuclear tests were conducted at Novaya Zemlya, including 87 atmospheric explosions (84 airbursts, one ground level, two water surface), three underwater and 42 underground.

The end of the Cold War and the demise of the Soviet Union did not end all nuclear testing at Novaya Zemlya. Russia's Ministry of Atomic Energy which succeeded the equivalent Soviet ministry and was itself in turn succeeded by Rosatom – that is Russia's atomic energy agency – undertook a number of subcritical nuclear tests there where no critical mass was reached and therefore no self-sustaining *nuclear* fission chain reaction occurred. In addition, the Arctic archipelago and the surrounding waters were, and remain, a depository for spent nuclear fuel as well as other nuclear waste.

Front view of the RDS-1. The two 'fish eyes' at the front were radar altimeters.

Kazakh Semipalatinsk, as well as Russia's Novaya Zemlya, were not the only locations of Soviet nuclear tests, but they were definitely the places where the largest number of tests were conducted and consequently also the ones which were most affected. The testing left a sorry legacy of radioactive waste, a landscape scarred by nuclear explosions and, last but not least, literally hundreds of thousands of people suffering adverse health effects. One may hope that this should suffice as a warning to the dangers presented by playing with nuclear fire but if experience with human nature is anything to go by, such hopes might be too great…

The Nuclear Forge

The tale of the test sites would be incomplete without devoting at least a few words to where the nuclear weapons were forged. In 1946 *Konstruktorskogo Byuro 11* (Design Bureau 11, abbreviated KB-11) was officially established, headed by Pavel Mikhaylovich Zernov with Yuliy Borisovich Khariton as chief designer. The main task of this organisation became the design and development of nuclear weapons. In 1947 it was decided that KB-11's facilities would be located in Sarov, a town in the Nizhny Novgorod oblast. One cannot help but to notice that Sarov was an important religious site of the Russian Orthodox Church and became a place where ways to figuratively open the gates of hell were concocted: to call this a devilish twist of fate would not be an exaggeration. Since, for obvious reasons, work on nuclear weapons was best done in secrecy, Sarov became a 'closed city': a location with official travel and residency restrictions. There were many such places in the Soviet Union and there are still some in the modern-day Russian Federation. In addition, the town was removed from non-classified maps and if mentioned at all was referred to by a number of official and unofficial code-names, the wieldiest used being Arzamas-16.

Nowadays, Sarov, the town having had its name restored upon the expressed wish of its residents, still remains a closed city. The reason for this being that the *Vserossiyskiy nauchno-issledovatelskiy institut eksperimentalnoy fiziki* (the All-Russian Scientific Research Institute of Experimental Physics)

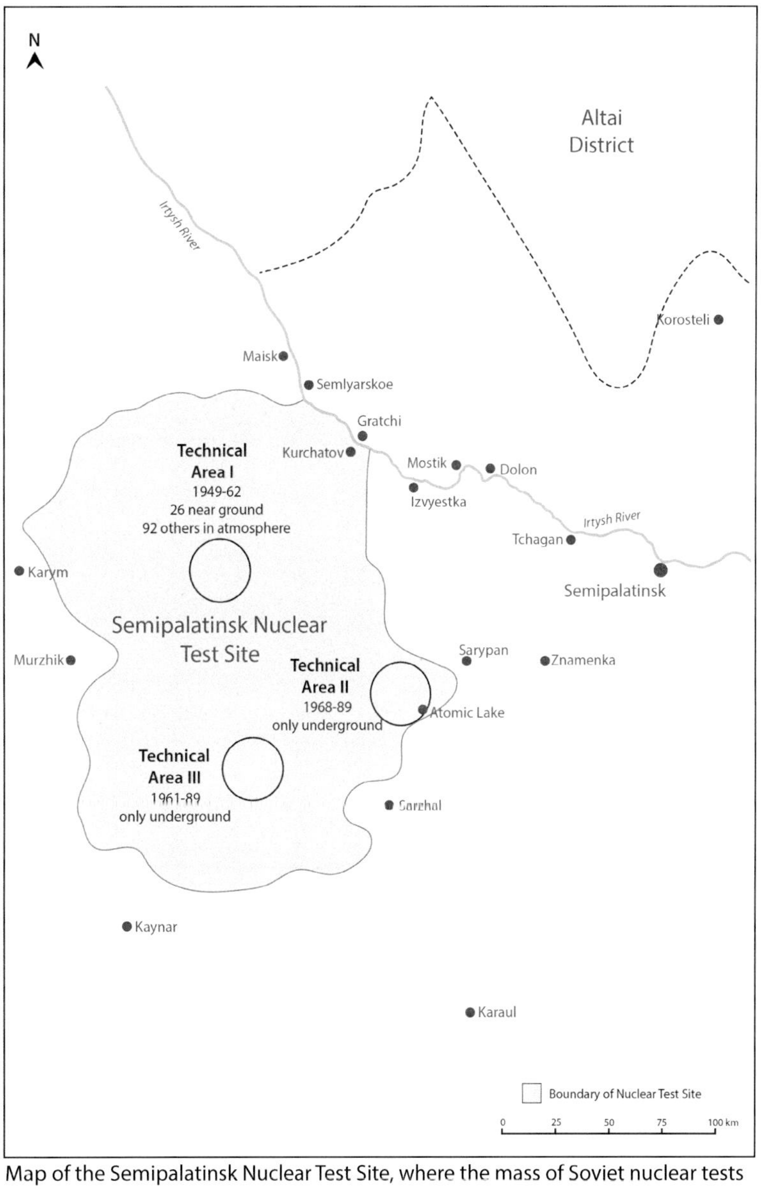

Map of the Semipalatinsk Nuclear Test Site, where the mass of Soviet nuclear tests was undertaken. (Map by George Anderson)

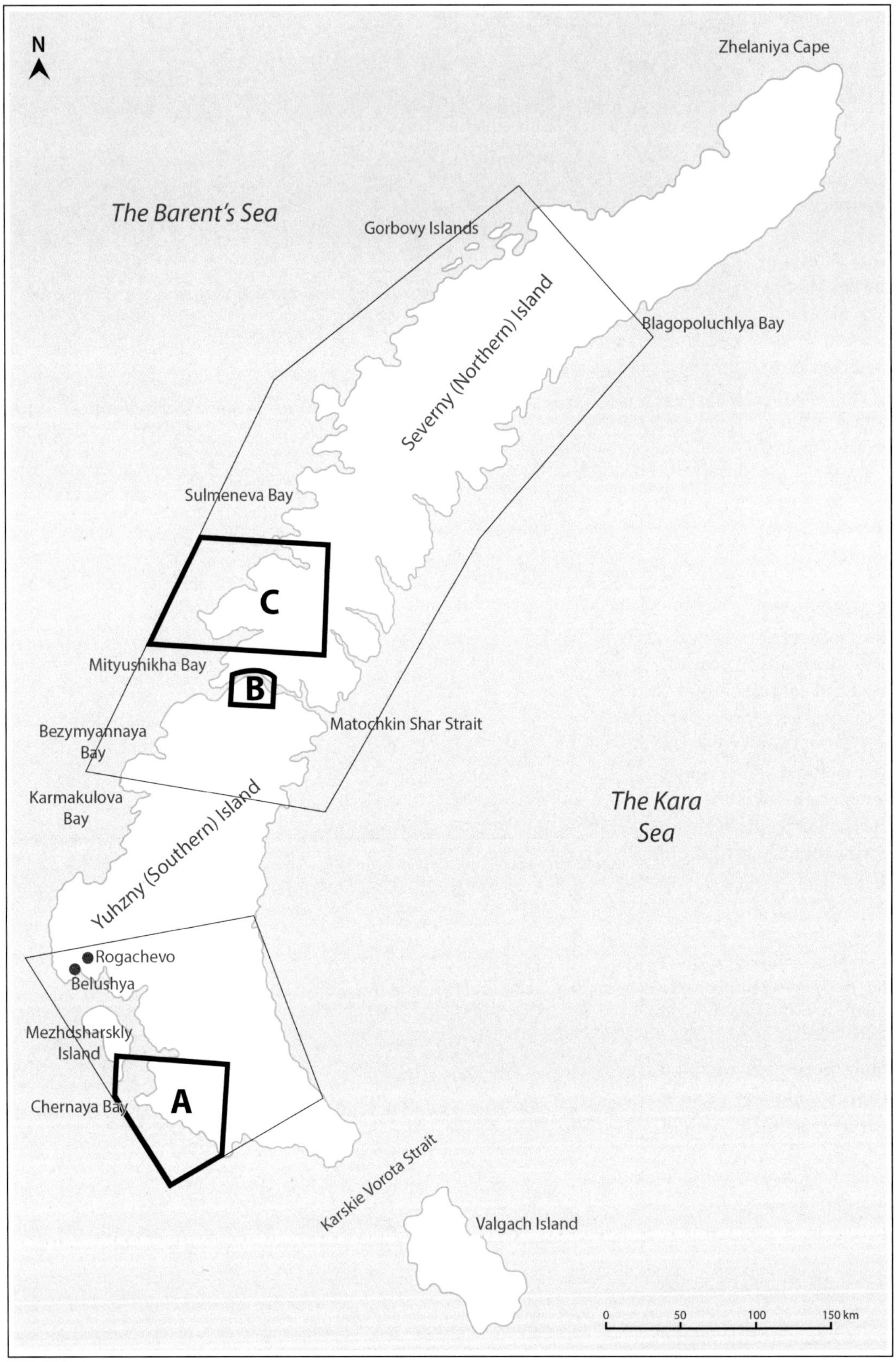

A map of the nuclear test sites at the Novaya Zemlya archipelago. (Map by George Anderson)

is headquartered there. However, its very existence is not a secret anymore, a point which is best illustrated by the fact that Sarov became a sister city of Los Alamos, New Mexico, a fitting match for reasons not difficult to deduce. Moreover, Sarov hosts the *Muzey yadernogo oruzhiya*, which is the Nuclear Weapons Museum, as well as the *Muzey-kvartira Yu. B. Kharitona*: Yu. B. Khariton's apartment-museum. The Nuclear Weapons Museum has an exhibit of aerial nuclear bomb and missile warhead casings, while the other facility also has many interesting pieces on display. However, since as already explained the museums are located in a closed city, accessing them may prove no easy task. Thus, the majority of those interested will have to satisfy their curiosity by viewing online galleries of the exhibits.

Much could still be written, but it should suffice to say that the nuclear age and the prospect of worldwide nuclear Armageddon which comes with it are not a thing of the past – the existence of closed cities being among the remainders of this fact.

A Suitable Carrier

As well as actually developing and fielding a nuclear bomb, there was the issue of the means of its delivery to the target. The Russians, and subsequently the Soviets, were among the first to develop and field heavy bombers in meaningful numbers. However, early Soviet aerial 'heavyweights' quickly fell into obsolesce, while due to a variety of factors efforts to develop more modern types in this category fell short of expectations. To make a long story short: by the later stages of Second World War when it came to heavy long-range bombers the Soviets lagged behind both in quality as well as quantity in comparison with their Western counterparts. In order to remedy this, Stalin more than once sought the delivery of American B-29 Superfortresses – the most advanced heavy long-range bombers of their time – yet all his requests to be supplied with them were turned down.

Nevertheless, the Soviets were to obtain the desired B-29s by other means. Namely while the Soviet Union was allied with Western powers against Germany it was also at the same time neutral with Japan (this would change almost at the very end of the war). That in turn meant that US aircraft operating against Japan which for whatever reason had to land in Soviet territory were interned. While the Soviets would in due time release their crews, they kept the aircraft for themselves. While the USAAF was conducting a bombing campaign against Japan three of the coveted B-29s were forced to land in the Soviet Far East:

- B-29-5-BW, serial number 42-6256
- B-29-15-BW, serial number 42-6365
- B-29-15-BW, serial number 42-6358

In addition, B-29A-1-BN number 42-9329 crashed in Soviet territory but many of its parts were recoverable.

Yuriy B. Khariton, one of the key figures of the Soviet nuclear weapons program, sitting next to 'his child', the casing of the RDS-1 nuclear bomb.

Naturally the question of how to utilise these 'gifts form they sky' arose, that is to incorporate their technology into prospective Soviet designs or copy the American aircraft and create its Soviet clone? Brushing aside any objections, Stalin personally opted for the latter – entrusting the project to Andrei Nikolayevich Tupolev and his design bureau. Accordingly, the B-29s were test flown and subsequently disassembled with each part being carefully studied. It took considerable time and effort to copy everything but many difficulties were encountered. For example, the B-29's skin was of 1/16th inch thickness: This would have been an impractical 1.5875 millimetre and therefore the skin of its Soviet copy was of between 0.8 and 1.8mm thickness, depending on the part of the aircraft it was covering. Other difficulties also had to be overcome and to deal with some of them Soviet agents in the West purchased items needed on the war surplus market. The first Soviet clones of the B-29 were ready by 1947 under the designation Tu-4. The Air Standardisation Coordinating Committee (ASCC) assigned it the designation 'Bull'.[7] Following extensive testing, the Soviets went ahead with the bomber's serial production, which by 1949 was in full swing. The aircraft were built until 1952, by when 847 had been delivered.

The availability of three original B-29s that landed in the USSR during the closing stages of the Second World War not only enabled Tupolev to develop a copy and launch series production of the Soviet equivalent but continued exercising strong influence upon Tupolev's designs for more than a decade longer.

The Tu-4 closely replicated the design of the B-29 but was not an exact copy. Apart from differing in skin thickness it was also decided not to build a Soviet version of the Wright R-3350 engines powering the American bomber. Instead, a variant of the M-71 engine, itself a clone of another Wright engine, was chosen. The resulting ASh-73TK power plant proved problematic at first but subsequent improvements made it fit for service. Similarly, the remote fire control system of the defensive armament was copied but American 0.5in (12.7mm) machine guns were replaced

A row of Tu-4s at a base of the Soviet Long-Range Aviation in the 1950s. Notable is the application of the construction number on the nose of each aircraft, followed by their two-digit 'Bort' number.

A top view of a Tu-4 in flight. Outwardly, the aircraft was almost completely identical to the B-29 except for its prominent national markings in the form of large red stars, outlined in white and red, of course.

Today a single Tu-4 is on display at the Monino museum in the vicinity of Moscow. Of the three American B-29s which landed in the Soviet Union and the one which crashed, nothing is known to have survived bar one item: a plaque reading 'At the request of the workers of the Boeing plant in Wichita, Kansas, this B-29 is named the General H.H. Arnold' – which was placed next to the bombardier's seat in one of the Superfortresses, and is believed to now be in private hands. Arguably it demonstrated to its new owners that back in those days the class divide in America might have been not as deep as Soviet propaganda had it portrayed.

The First Drop

Just exploding a nuclear bomb without being able to deliver it upon a target would be nothing but an exercise in futility. Before missile technology was sufficiently developed the best means of delivery was offered by a bomber. Thus, the Soviets proceeded to gather data regarding the effects of a nuclear explosion on an aircraft in flight. To that end, during their second nuclear test on 24 September 1951, an overflight by a fully crewed Tupolev Tu-4 bomber with Captain Konstantin Ivanovich Usachev at the controls was performed. The RDS-2 (Joe-2) nuclear bomb of 38kt yield undergoing the test was placed on a 30m high tower (positioned at the coordinates 50.43°N and 77.83°E) while the aircraft flew at an altitude of 10,000m. The bomb's detonating mechanism was set into action via a radio signal as the aircraft passed over the tower, with the nuclear explosion itself occurring some 1.5 minutes thereafter. The shockwave of the nuclear blast caught up with the bomber as it was about 20 kilometres from the point of detonation. The test demonstrated that the effects of the nuclear explosion were nothing the aircraft and its crew could not handle.[8]

with Soviet 23mm NS-23 cannons. In addition, once in production and service the Soviets went ahead with various modifications and special versions (for example a variant carrying anti-ship guided missiles) which further 'branched off' the Tu-4 from its B-29 'root', yet even with all this the Soviet bomber's American ancestry was immediately recognisable.

By the early 1950s, and not long after entering service, the Tu-4 was becoming obsolete. Newer jet powered bomber designs were superseding it in virtually all characteristics. Moreover, combat experience during the Korean War clearly showed that while American B-29s could rain down destruction upon the enemy below, they also had vulnerabilities: ironically Soviet MiG-15s demonstrated that the B-29s were particularly vulnerable to them, in turn meaning that the Tu-4 would fare the same, if not worse, when facing American or other Western jet fighters. Yet by that time the Tu-4 had already played its part in giving the Soviets a heavy long-range bomber, one that was also adapted for the nuclear strike role it should be noted. In addition, it plugged a gap in Soviet aviation technology development and provided Tupolev with a solid base for follow-on design work of modern heavy long-range bombers.

Since it was empirically proven that a nuclear blast would be survivable for an aircraft and its crew the ultimate test of actually dropping a live nuclear bomb from a bomber became the next logical step. This was scheduled for 18 October 1951; on that day the third Soviet nuclear bomb, accordingly designated RDS-3 (Joe-3), was to be dropped from a bomber at the Semipalatinsk Test Site in Kazakhstan. The testing itself, that is airdropping and detonating the bomb, was the crowning of a process which was both time-consuming as well as laborious. However, the immediate preparations of the aircraft for the test were not particularly long, beginning a day before the scheduled date of the test. Being the first Soviet test of a nuclear bomb dropped by an aircraft meant that much was done for the first time and was therefore novel. Thus, not surprisingly, almost every procedure during the test itself, as well as the preparations for it, in some ways deviated from the established routine. Even arming the aircraft with the bomb had to be conducted in an unusual way, and not only due to the war load's unconventional nature. Because of the dimensions of the RDS-3, and the Tu-4's clearance, it was not possible to just load the bomb like any other piece of heavy ordnance. It proved necessary to prepare an opening in the ground in which the bomb was placed and over which

Another long row of Tu-4s at an air base, seen in the late 1950s. Some of these bombers were subsequently sold to China, where they served as the first strategic bombers of the People's Liberation Army Air Force, and then as testbeds for various purposes.

Bottom view of a Tu-4 overflying its air base at a relatively low altitude. A total of 847 Tu-4s had been built by 1952, many in the form of experimental variants including a fuel-transport and a tanker variant, radiation reconnaissance aircraft, communication relay aircraft and a remotely controlled target drone.

the bomber was then positioned. When this was appropriately arranged, the RDS-3 was hoisted into the Tupolev's bomb bay and secured, with the release and other mechanisms then connected. Subsequently all the necessary checks of the aircraft, its load and associated systems were performed and once completed readiness for the mission was reported.

When, finally, the go-ahead was given the bomber's crew under the command of the first pilot HSU Lieutenant-Colonel Konstantin Isaakovich Urzhuntsev boarded the aircraft and manned their stations. Having received permissions for take off the Tupolev took to the air from the runway of the Zhana Semey (Jana-Semey) airbase – which is located about eight kilometres south of the city of Semipalatinsk – climbed to an altitude of 10,000 metres and set course for the target. The aircraft and its systems performed without any malfunction and despite the understandable tense anticipation, the flight to the target was practically eventless. Thanks to good visibility, making out the test site did not present a problem and the bomb was released over the intended target. As the nuclear bomb first freely hurtled and then drifted down under a parachute, its detonation sequence was initiated via a radio signal with the exact timing being determined by a barometric sensor so as to detonate at the optimum altitude. On 18 October 1951, at 9.53hrs Moscow time, the explosion which had a 42kt yield occurred at an altitude of 380 metres above ground. The coordinates 50.43°N and 77.83°E were recorded, which was in the vicinity of the two previous Soviet nuclear tests. Meanwhile, anticipating the flash as well as shock wave of the nuclear blast, the bomber's crewmen pulled darkened goggles over their eyes and tightened their seatbelts. Indeed, the shockwave caught up with the aircraft, shaking it somewhat, but the crew experienced no problems in maintaining control. On the return leg everything worked as it should and the Tupolev touched down without incident. Having landed, the crew underwent a decontamination of sorts as they discarded all the clothing worn during the test, took a bath and changed into freshly provided uniforms. After performing, arguably, the most spectacular and exciting mission of their careers, the crewmen now set about the mundane task of preparing their post-flight reports.

Before moving on, it is worth mentioning that the nuclear bomb-carrying aircraft was accompanied by a 'double' which was also a Tu-4 – flown by the same men as during the RDS-2 test – with an inert FAB-1500 in its bomb bay, tasked to guide the main aircraft in the event that due to a failure of its navigational or bomb aiming equipment – or for any other reasons – it could not locate the target. Should that have occurred, the 'double' was to assume the lead with the nuclear bomb-armed Tupolev following in its wake. Once over the designated target the 'double' was to signal the RDS-3-carrying aircraft by radio as well as drop its inert FAB-1500 bomb. At that time the Tu-4 with the nuclear bomb was also to release its load. However, it transpired that there was no need for the arrangement described as the nuclear-armed aircraft did not experience any significant problems. In addition, aerial cover for the entire undertaking was provided by Lavochkin La-11 fighters, yet this precautionary measure proved to be unnecessary as nothing warranting their intervention occurred during the test.

Summing it up, it can be stated that the test was almost surprisingly trouble-free, with the aircraft, the various systems, and the bomb itself performing without any significant malfunction and last but not least the men also proved to be up to the task. The significance of the successful conduct of the test lay in that from this day onwards the Soviet Union became an actual – as opposed to a nominal – nuclear power, for it not only had tested a functional nuclear device, which was a prerequisite, but also successfully tested the means of its delivery.

The Men Who Dropped the Big One

Nuclear bombs were dropped *from* aircraft but, they were dropped *by* men, who also – let it not be forgotten – flew the said aircraft in the

Lieutenant-Colonel Urzhuntsev, pilot and commander of the Tu-4 that performed the first live nuclear bomb drop in the USSR.

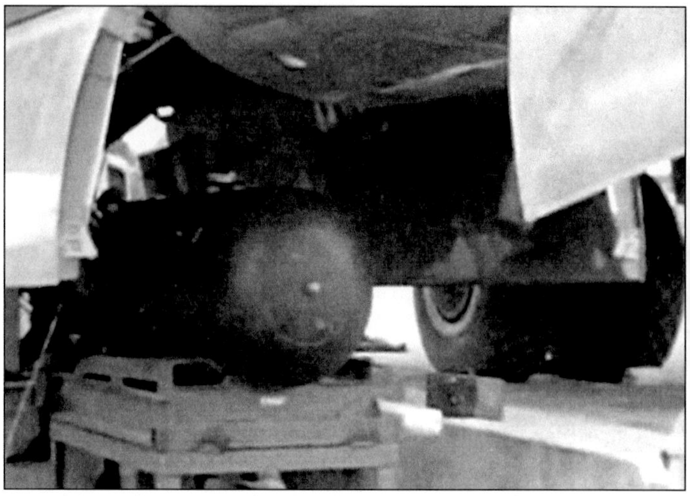

One of the early Soviet nuclear bombs, in the process of being hoisted into the bomb bay of a Tu-4 bomber.

first place. Without these men's skill, and the will to utilise the aircraft and the bombs they carried, both would have for all practical purpose remained useless. Yet in spite of this being seemingly obvious, the airmen who flew the bombers and dropped nuclear bombs remain by and large anonymous. The observation made in particular applies to Soviet airmen who – bar a few exceptions – are virtually unknown in the English-speaking world. This remains so despite the fact that for a long time their identities have not been a secret, as crew lists of Soviet bombers are not only accessible in the archives but have been published in the traditional form and online. Without over analysing the issue raised, it is possible to make up for the deficiency by providing herein the crew list of the Tu-4 which dropped the RDS-3 on 18 October 1951 and of the Tupolev which participated in the 24 September 1951 test of the RDS-2 device and subsequently acted as the 'double' during the October airdrop. The aircraft were fully manned as if for a wartime combat mission.

The crew of the first aircraft were:

- Hero of the Soviet Union Lieutenant-Colonel Konstantin Isaakovich Urzhuntsev: first pilot and commander
- Senior Lieutenant Ivan Mikhaylovich Koshkarov: second pilot
- Captain Vladimir Semenovich Suvorov: navigation
- Captain Boris Dmitriyevich Davydov and Senior Lieutenant Kiryushkin Nikolay Dmitriyevich: additional navigators who also fulfilled the duties of the bombardier and systems operator respectively
- Major Vasiliy Nikolayevich Trofimov: flight engineer
- Lieutenant Arkadiy Fodorovich Kuzniecov: flight technician and assistant to the flight engineer
- Junior Lieutenant Vladimir Vladimirovich Yakovlev: communications
- Private Arkadiy Fedorovich Yevgodashin: defensive armament operator
- Lieutenant Alvian Nikolayevich Stebelkov who was entrusted with handling the bomb itself.

Before moving on to the crew of the second Tupolev it seems fitting to elaborate on the Hero of the Soviet Union title of Lieutenant-Colonel Urzhuntsev who held the title and corresponding medal no. 8313. This was awarded for his wartime service with the decree bestowing the title upon him being promulgated on 23 February 1948.[9] As for the ranks and professional titles of the airmen listed they are a direct translation form the Russian language which has the disadvantage of sounding somewhat clumsily when worded in English but make for this up by being true or at the very least close to the original.

The 'double' was similarly crewed by:

The crew of a Tu-4 receiving a pre-flight briefing before a training sortie.

- Captain Konstantin Ivanovich Usachev: first pilot and aircraft commander
- Senior Lieutenant Vasiliĭ Ignatovich Kureev: second pilot
- Captain Aleksey Alekseyevich Pastunin: navigator
- Senior Lieutenant Georgiy Alekseyevich Sablin: bombardier
- Senior Lieutenant Nikita Ivanovich Svechnikov: system operator
- Lieutenant Petr Porfiryevich Cherepanov: Flight engineer
- Lieutenant Filaret Ivanovich Zolotukhin: flight technician
- Senior Sergeant Vladimir Borisovich Zolotarev: communication
- Senior Sergeant Nikolay Dmitriyevich Borzdov: defensive armament
- Lieutenant Leonid Andreyevich Blagov: held the somewhat awkwardly sounding title of trials technician.

RDS-4 – or Natasha – was the first 'miniaturised' Soviet nuclear bomb. While made to fit into the bomb bay of the Il-28 light bomber, and approximately one third the size of the RDS-3, it was still 3.365m long, weighed 1,200kg (2,646lbs), and had a yield of 28kt.

An Il-28 like the example that dropped the Natasha, even having the same 'bort' number (for details, see the colour section). Oddly, the Il-28 had an unswept wing design, and was powered by a Soviet version of the Jumo 004 turbojets of Second World War vintage. Nevertheless, it easily beat the rival Tu-78 prototype and then proved such a success in operations that the basic design was hardly changed in its career. (Tom Cooper Collection)

A Close Second

The first successful airdrop of a nuclear weapon was not the apex of Soviet efforts to practically test nuclear aerial bombs, but merely its beginning. It needs to be pointed out that the weapon employed was a bulky piece of ordnance, while the delivery platform was a propeller driven aircraft – both of which were by the early 1950s on the verge of obsolescence. Rapid technological progress in aviation technology – such as jet powered flight – as well as new delivery methods by means of missiles, which were also being developed, called for matching advances in nuclear weapons technology. Among other types, the Soviets fielded the Il-28 jet powered twin-engine light bomber and were also working on surface-to-surface missiles. In order to give them the capability to deliver nuclear payloads, new types of nuclear ordnance had to be developed which would have smaller dimensions and be of lighter weight than the weapons already available. The Soviets managed to build such a weapon in the form of the RDS-4 nuclear bomb, colloquially known as 'Natasha'. Once the new weapon was available testing it became imperative.

Before carrying out the live test a rehearsal was judged to be necessary. Thus, on 18 March 1953 an Il-28 bomber test-dropped a practice bomb which was a perfect imitation of the actual one, except that it lacked the nuclear warhead and carried telemetric equipment instead. By the summer all data gathered was analysed and any arising issues had been delt with. Hence, all was ready for test dropping a live RDS-4 nuclear bomb which was to take place on 23 August 1953. The nuclear bomb was to be delivered by an Il-28 crewed by V. I. Shapovalov (pilot), A. V. Kozminykh (navigator) and B. S. Sudakov (radio operator/gunner). The first aircraft was accompanied by a 'double' in the form of another Il-28 with the crew positions taken by M. I Loshakov, A. M. Karmin and I. N. Sharkov respectively. The two Ilyushin bombers were to be escorted by a pair of MiG-17 fighters. On the scheduled day the formation took to the air heading for the target located at the Semipalatinsk Test Site. The RDS-4 nuclear bomb was released from an altitude of 11,000 metres. Set for an airburst it detonated at an altitude of 600 metres at the coordinates 50.4°N and 77.8°E. The explosion was estimated to have had a yield of 28kt. The available sources are rather laconic when describing the test, providing little information bar the basic facts. However, it can be concluded that it was a success as apparently no significant problems were reported. Natasha's test became known as Joe-5 in the West.

While all attention is usually focused on an event which is first in any given category, the second such is more often than not relegated to obscurity and RDS-4's test was not an exception. Yet while it was indeed the second Soviet airdrop test of a nuclear bomb it was also a first in two significant categories; namely the first such from a jet powered aircraft; and the first test of a tactical nuclear weapon. Both of these were important as the Soviet jet powered twin-engine light bomber was provided with a tangible nuclear capability. Moreover, the RDS-4 – even though the weapon itself remained in service only for a relatively short time – was a major advance in Soviet nuclear

The Bomber Gap

The first post-Second World War modern Soviet heavy bomber was a 'clone' of the American B-29 which was designated Tu-4. Several hundred were built, of which a number were configured for nuclear bomb delivery. Yet already by the mid-1950s the Tu-4 was obsolete. That seemed not to be a particularly serious problem as new heavy bombers were either entering service or getting close to it. The Tupolev design bureau could pride itself with two new creations: the Tu-16 jet bomber and the Tu-95 turboprop one. If longevity is anything to go by the latter proved most successful. In addition, a new design bureau headed by Vladimir Mikhaylovich Myasishchev presented its creation; the M-4 bomber; soon followed by an improved variant the 3M. Myasishchev's aircraft were designed to have intercontinental range and thus to be truly 'strategic', for it meant they would possess the capability to reach targets in continental North America.

Rather unsurprisingly, these developments were closely followed in the United States. When evaluating Soviet bombers US analysts compared the Tu-16 to the B-47 Stratojet and the M-4 to the B-52 Stratofortress. The Myasishchev bomber in particular caused much alarm. It was first revealed during a flypast at the annual May Day parade (International Worker's Day – lavishly celebrated in the Soviet Union for ideological reasons) in 1954. Subsequently it was displayed at a number of other flypasts and parades. In particular the flypast during the Soviet Aviation Day at the Tushino Airfield in July 1955 was to have far reaching consequences. During this event 10 M-4s were flown in view of the public: the formation then flew out of sight and joined up with eight additional bombers before turning around and then making another pass. This resulted in the illusion that the Soviets had shown no less than 28 bombers in a single fly-by – the initial formation of 10 and a second of 18. Seeing 28 Myasishchev bombers, Western analysts came to the conclusion that by 1960 the Soviets would have 800 of these in operational service.

Reality proved different to the analytic estimates. It was not known in the West at that time that Myasishchev bombers were difficult to fly, were unreliable as they were technically troublesome, could not be armed with some of the bulkier pieces of nuclear ordnance due to their bicycle undercarriage, and – most importantly – their unrefuelled range was actually insufficient to reach the USA. For these reasons – and in particular the latter – it was decided not to pursue their large-scale production. Certainly enough, and together with Tu-95s, they *did* represent the first Soviet delivery systems capable of reaching US territory, and, from the moment they entered service in 1956, until the early 1960s, they were the only Soviet intercontinental delivery systems. Meanwhile, aerial reconnaissance by U-2 aircraft, which penetrated deep over the Soviet Union, showed that only a limited number of Myasishchev bombers were deployed at operational air bases, and there were no signs of large-scale production. However, the results of these early U-2 missions were kept 'top secret'. Moreover, some of those who had access to the reconnaissance material gathered, in particular the

A classic Cold War-era photograph showing a B-52D – the primary nuclear-armed bomber of the Strategic Air Command – in the process of in-flight refuelling from a KC-135A tanker. With help of the tanker fleet, SAC was capable of reaching virtually every corner of the USSR from any air base in the world. (US DoD)

The development of the Tu-16 was motivated by the need to develop a high-speed jet bomber in theatre missions: indeed, one of their primary aims was to double the Tu-4's speed and thus enable the bomber to evade enemy jet fighters.

The first Tu-16s entered service in early 1954 and began replacing Tu-4s. They were followed by Tu-16As, developed specifically to deliver nuclear weapons. Visible in this photograph, this version included an insulated bomb bay and undersurfaces painted in white to protect them from nuclear radiation.

The first Tu-95s entered service in 1956, and initially also served as replacements for the Tu-4s. Unknown at the time was that – thanks to its reliability and versatility – ultimately, the type was to outlast all other Soviet bombers in service in the 1950s.

A semi-profile of the Myasishchev M-4 Molot ('Hammer') bomber (ASCC/NATO-codename 'Bison A') serial number 4300205. The first prototype of this aircraft flew on 20 January 1953: state acceptance trials began in March 1954, and production was launched later the same year, resulting in service entry in 1955. During the testing, an M-4 carrying a 5,000kg load reached a range of 8,500 kilometres: this was still insufficient to reach the USA.

Front view of an early M-4, an aircraft that caused a big surprise for the USA when displayed to the public for the first time on 1 May 1954. Although faster and carrying a greater payload than the famous Tupolev Tu-95, Myasishchev's design proved difficult to fly and technically unreliable. Soon after production was launched, attempts were made to install the Kh-20 cruise missile on the M-4, but these were abandoned due to the low clearance of the missile under its fuselage. Consequently, only 34 examples (including two prototypes) were ever built.

A top view of a Myasishchev 3MD (ASCC/NATO-codename 'Bison C') intercepted by Western fighters over the North Atlantic in 1968. This version was powered by more fuel efficient engines, and was equipped with an in-flight refuelling probe to increase range. While originally built as cruise missile-carriers, most 3MDs were converted to the tanker configuration during the late 1950s.

senior leadership of the USAF, had a vested interest in maintaining the image of a Soviet bomber threat as it provided them with a convenient justification for demanding a substantial build-up of their own bomber force.

As things stood – without a realistic picture of Soviet bomber capabilities being widely available – a Soviet 'superiority' in bombers vis-à-vis the US became known as the 'bomber gap'. This was to have far reaching consequences as it resulted in a number of American offensive and defensive countermeasures. The United States Air Force's Strategic Air Command would be substantially built up, receiving thousands of Boeing B-47 Stratojet and Boeing B-52 Stratofortress strategic bombers, supported by hundreds of Boeing KC-135 Stratotankers.

With this the United States for some time gained a decisive edge in the number of strategic nuclear weapon delivery systems over the Soviet Union. This advantage would only be overcome by the Soviet intercontinental ballistic missile arsenal, but its development and build-up were still many years away. Meanwhile, in the defensive realm the feared Soviet bomber threat led to the creation of the North American Air Defence Command and its massive machinery, including hundreds of manned interceptors, surface-to-air missiles and related bases, large early warning radar stations, and the development of early warning aircraft.

One could endlessly argue on how justified and necessary the American response was, however, it should be taken into consideration that in view of publicly available information at that time a substantially different reaction would have hardly been possible. Moreover, even though the Soviets were in fact able to carry out only very limited attacks against North America, one could ask: how much would have been acceptable? One, two or three thermonuclear bombs? Hundreds of thousands, a million or more casualties? Even a very few thermonuclear bombs on major urban centres could result in such numbers of killed and wounded. Thus, building up a retaliatory – as well as a defensive – capability was not completely out of place even if the 'bomber gap' never existed.

Right-side profile of an M-4 about to touch down. Its potential advantages in comparison to the Tu-95 prompted the Soviets to continue improving the original design. In 1956, the 3M – incorporating a major redesign of the nose and the centre fuselage (both aimed at reducing weight and improving aerodynamics) flew for the first time.

weapon technology. Its compact design allowed not only for carriage in the weapons bay of light bombers but also to serve as the basis for a nuclear warhead for surface-to-surface missiles. In particular the warhead of the R5M (8K51) ballistic missile which was assigned the ASCC reporting name SS-3 Shyster was derived from the RDS-4 nuclear bomb. To summarise it in a succinct manner: the 23 August 1953 Soviet nuclear test was their second test of an airdropped nuclear bomb but it was a close second in terms of the significance for the Soviet nuclear weapons development program.

Unsung Heroes

The importance of gathering data during live tests of nuclear weapons was self-evident, but it was equally obvious that this would be hazardous due to the danger presented by radiation. In order to deal with this problem, it was originally intended to utilise unmanned means to gather the data needed. For this purpose, Yakovlev Yak-9V fighters were modified for remote radio control from an appropriately fitted Tupolev Tu-2 'mothership'. While the combination was technically workable, operating it presented a much bigger problem and it proved particularly difficult to land the unmanned radio-controlled aircraft without crashing them. For this reason the program was abandoned. Perhaps if the Soviets had been willing to sacrifice a few more Yaks and some additional time for the operators to master the nuances of remote piloting, Soviet unmanned aviation would have attained a level it never actually reached. However due to the difficulties encountered such then-unorthodox solutions as remotely controlled aircraft were abandoned in favour of manned ones.

As things stood a number of manned aircraft types were utilised to gather data from nuclear explosions, such as propeller driven Li-2 transports. However due to the need to keep up with more modern and faster jet and turboprop bombers modified Il-28s and Tu-16s were adopted. It was only with the introduction of the Il-28 in this role that effective protection for the crew could be given. Namely in then-modern jets the cockpit could be over-pressurised, preventing outside air containing radioactive particles from entering and thus contaminating it.

Initially the equipment fit of the aircraft concerned was modest, if not outright rudimentary, as only tubular filters were carried in order to gather radioactive particles from the air. These were designated *samolety radiatsionnoy razvedki*, or radiological reconnaissance aircraft – which fittingly described their mission. However, soon the set of sensors carried would be considerably enhanced, comprising a whole range of instruments meant to measure radiation levels, air pressure, light intensity as well as other data. In addition, high resolution cameras would also be carried so as to film the tests, including the bomb drop and the explosion. This was necessary in order to gather vital data on ordnance release and separation fromm the aircraft dropping it, the in-flight ballistic characteristics of the ordnance, and last but not least the nuclear explosion itself, that is the forming of the fireball and the promulgation of the shockwave. The aircraft thus modified were termed *letayushchaya laboratoriya* or *samolet laboratoriya*, meaning flying laboratories or aircraft laboratories. While all the attention usually went to the aircraft and men who test-dropped the nuclear bombs, the missions of the radiological reconnaissance aircraft and the flying laboratories were both indispensable in nuclear testing – as well as hazardous – making the airmen who flew them the unsung heroes of nuclear weapons development.

2
PLAYING NUCLEAR WAR

Having tested nuclear weapons and the means to deliver them made the Soviet Union a true nuclear power. However, it was thought that merely dropping and detonating such devices did not in fact give a clear enough picture of how an actual nuclear war might look in practice. Short of starting one, the closest to 'the real thing' would be a large-scale military exercise involving a substantial number of troops and military equipment, and one which would also involve the detonation of an actual nuclear weapon. Thus, the Soviet leadership decided to literally play nuclear war, but unlike children who only have plastic soldiers at their disposal to toy with, they would expose living men to the effects of a nuclear weapon.

Lightning in the Urals

The decision to conduct military manoeuvres with live nuclear weapons had to be taken at the highest leadership level. Accordingly, a decree to that end was promulgated by the Central Committee of the Communist Party of the Soviet Union and the Council of Ministers of the USSR on 29 September 1953, followed by an order of the Soviet Minister of Defence dated 5 November 1953. To sum it up, the basic premise of the exercise was to simulate an offensive operation with the use of nuclear weapons against an enemy who took up defensive positions. Considering the realities of the time period in question there was little doubt who the enemy would be. Thus, it should be no surprise that the Soviets were looking for a location with terrain features resembling those which could be encountered in Western Europe and found this in Orenburg Oblast. There – in the vicinity of Totskoye village to be more precise – the war games were to be conducted with Marshal of the Soviet Union Georgy Zhukov in overall command.[1]

The basic premise of the manoeuvres – which were given the codename '*Snezhok*' (meaning light snow)(the nuclear test received the Western codename Joe-8) was an offensive operation with nuclear weapons being utilised against the enemy. In fact, the war game's scenario envisioned the use of three nuclear weapons, of which only one would be actually dropped and detonated, with the other two being simulated by large conventional bombs. While the nuclear strikes – the one involving a live nuclear weapon in particular – were the centrepiece of the whole undertaking it also involved large numbers of conventional forces: 45,000 officers and men of the Soviet Army would have 600 tanks and self-propelled guns, about the same number of other armoured vehicles, 500 artillery pieces of varying type and calibre as well as 6,000 trucks, cars and tractors. The main thrust of this corps-sized grouping of ground forces would be the carried by the 270th Rifle Division, while air support would be provided by numerous bombers and fighters with some of the latter strafing (more on this below).

Considerable preparations for the exercise were carried out on site. Many kilometres of trenches, dugouts and other earthworks were

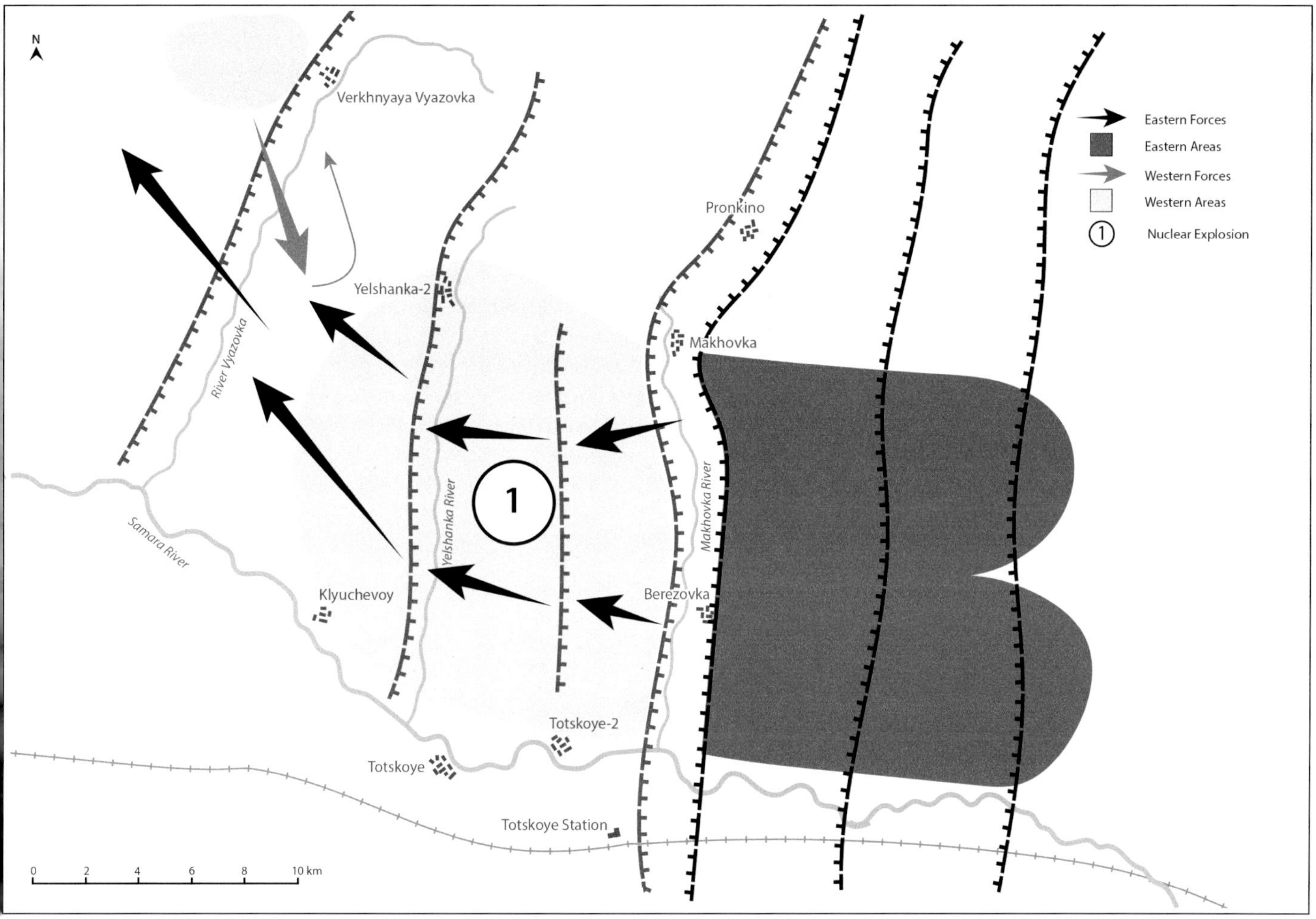

A map of the plan for Exercise Snezhok. (Map by George Anderson)

constructed both for Soviet troops as well as to simulate 'enemy' positions. In order to gather data on the effect of a nuclear explosion on living organisms, farm animals such as horses, cows and sheep were placed at various distances form the intended point of the explosion and with varying protection, such as being tied in the open or placed in trenches. Likewise, numerous pieces of military equipment, including aircraft, were also positioned so as to test the effects of a nuclear explosion upon them. Last but not least, the safety of the local population also had to be addressed. Those living within a radius of up to 8km from the intended point of explosion were evacuated, while people residing further away were for the most part left but advised to vacate their homes, take cover or lay flat and face down.

The date chosen for the exercise was 14 September 1954, this having several reasons. Amongst others by that day all necessary preparations would be finished. If the forecasts were anything to go by the weather would also be favourable and since the nuclear weapon was to be airdropped, particular attention was given to this aspect of the undertaking. At the Vladimirovka airbase in the Astrakhan Oblast a special aviation group was assembled under the command of General V. A. Tsjernorez, composed of two Tu-4 bombers, commanded by V.YA.

After being brought by train to Bagerovo Air Base on the Crimenan Peninsula – the home base of the 71st Special Polygon (the unit operating the Tu-4s involved in all of the Soviet nuclear tests) – the RDS-3 bomb was cautiously unloaded on its transportation trolley.

Kutyrcheva and K.K. Lyasnikov (one of which was to drop a nuclear bomb) and two Il-28 bombers tasked with observing the test including filming and photographing it. In addition, the group included six MiG-17 fighters to provide escort as well as a Li-2 transport and a Yak-12 liaison aircraft. Because of the distance between Vladimirovka and Totskoye, a single pair of fighters could not provide continuous escort for the nuclear-armed bomber. For this reason, six MiGs had to be used with fighters meant to relieve those running short of fuel based at Sorochinsk.

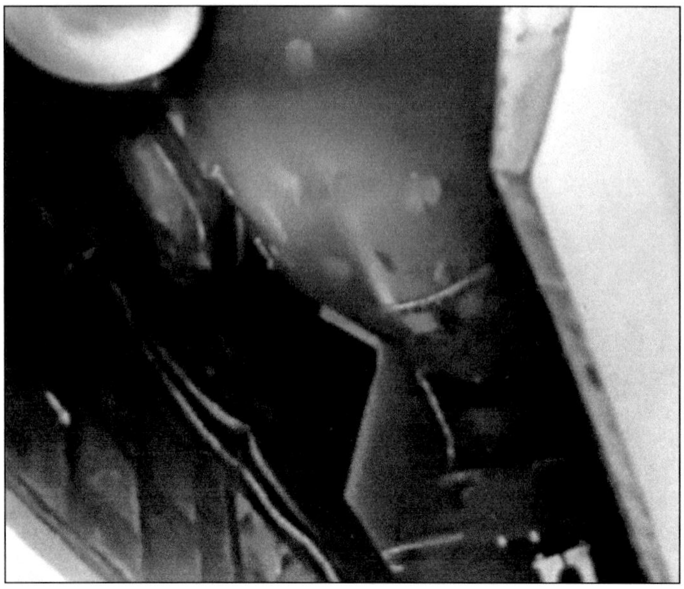

A last look at the RDS-3 bomb deployed during Exercise Snezhok, seen inside the bomb bay of the Tu-4 that deployed it.

Training flights for the mission started on 12 August. Initially the airmen were not aware of the precise nature other that it will involve the dropping of a nuclear bomb. They were to find out more just a few days before the planned date of the exercise when Igor Kurchatov visited the airbase to see for himself how the preparations were going and to talk with the airmen. It was only then that the crews were informed that they would not only drop a live nuclear weapon but that friendly troops would be close to its point of detonation. Learning this responsibility the airmen felt burdened. Meanwhile, training proceeded, including the dropping of practice bombs. Impacts were registered about 50-60 metres from the target, which were deemed to be good results. Thus, it seemed everything would go according to plan.

Finally, 14 September 1954 came. The weather forecasted was favourable with only weak winds and few clouds. Early in the morning the command to man the aircraft was given. A small aerial armada compromising two Tu-4 bombers (one of which was armed with a nuclear bomb), two Il-28 jets which were to observe as well as record the ongoing events, and finally two escorting MiG-17 fighters prepared to take off. By 06.00hrs (the time 06.28 also being given) the bomber with the nuclear weapon was in the air and shortly thereafter the second also took off. All the aircraft set course for the target in Orenburg Oblast. The target aiming point was clearly marked on the ground with a 150m rectangle which had a cross in its centre, supplemented by chevrons at a distance providing additional visual reference – all these markings being in white. Navigational aids were also placed along the ingress route starting at 72km from the target. All this was necessary as the nearest troops were only 5km away from the target aiming point and thus the Soviets could not afford to miss or the result could be an unimaginable disaster.

A pair of Tu-4s involved in Exercise Snezhok.

The Tu-4 bomber with its deadly payload on the way to the target. The crew was informed about the live drop days in advance, but friendly troops deployed in its immediate vicinity received similar information only immediately before the exercise was initiated.

By 09.20, when the bomber was already nearby, the weather in the target area started to unexpectedly somewhat deteriorate. In particular the speed of the wind, an important factor when dropping bombs, increased and was recorded as 20m/s. Meanwhile the Tu-4 with the nuclear bomb reached the target. At 09.34 navigator-bombardier V. V. Babiec released its deadly cargo from an altitude of 8,000 metres. Slowed down in its descent by a parachute, the bomb detonated 48

The aiming point for Exercise Snezhok was clearly marked on the ground by a 150m wide and deep rectangle, with a cross in the centre, supplemented by chevrons at a distance (one of which is barely visible in this image to the left of the rectangle). Due to the deployment of live troops, an accurate drop was of crucial importance, otherwise the exercise could have easily degenerated into an unimaginable disaster.

30 MiG-15 fighters strafed ground targets following a bombardment from Il-28 bombers escorted by MiG-17 Fighters.

Exercise Snezhok was a very comprehensive enterprise, including not only the dropping of a live nuclear bomb, but also the deployment of various conventional weapons. Visible here is a formation of Il-28 bombers, about to deliver a convention bombardment immediately after the nuclear strike.

seconds after being dropped at an altitude of 350 metres. Unfortunately, the bomb exploded some 280m in the north-western direction off the intended target aiming point (the coordinates 52.64418°N and 52.80547°E are given), though this did not compromise the safety of the troops on the ground.

Before that – 10 minutes to be precise – the code-word 'Molniya' (Lightning) was transmitted over the radio, this being the signal to take cover as a nuclear explosion was imminent. Thus warned, the troops huddled in the trenches and dugouts, lay flat on the ground next to artillery pieces and other crew served weapons which they manned or sat still in buttoned-down armoured vehicles. Minutes seemed to be hours but finally a blinding white flash literally eclipsed the sun as the bomb exploded. The bright flash in no time turned into a fireball and subsequently an ominous mushroom cloud formed. The shockwave could be felt, the earth trembled and the sound of a loud thunder, somewhat longer than a natural one, filled the air. As the earth shook, with the effects of the nuclear explosion being seen, felt and heard, the officers and men positioned in the exercise area could not escape the awe of the power of the weapon used. Meanwhile the explosion's effects were also experienced in the aircraft which dropped the bomb and when the shockwave caught up with the Tupolev it was lifted by some 50 to 60 metres, though the bomber did not sustain any serious damage.

While detonating the nuclear bomb could make for a dramatic climax, it was just the beginning from a tactical point of view. Five minutes after the nuclear blast a preparatory artillery barrage commenced – in fact two barrages each lasting 25 minutes. Numerous shells, rockets and mortar bombs were fired onto 'enemy' positions. Meanwhile a large formation composed of 86 Il-28 bombers escorted by 42 MiG-17 fighters appeared overhead at an altitude of 5,000 metres. Armed with 250kg bombs, the Ilyushins were to deliver a conventional aerial bombardment and in all dropped 688 bombs, of which 583 were said to have impacted 'enemy' positions. According to a post-exercise evaluation about 44 percent of 'enemy' positions thus treated were 'destroyed' with corresponding 'enemy' casualties estimated as amounting to 25 percent. While performing the mission, because of the need to maintain formation, 39 bombers and six fighters had to fly through the mushroom cloud. However, this had no

A disused Lavochkin La-15 fighter about to be subjected to the destructive effects of a nuclear blast. Lavochkin's creation was superseded by the MiG-15 and withdrawn from service after a relatively short production run, which is why airframes like this one were then spent as targets.

The nuclear explosion blew a T-34 tank's turret clean from the hull, providing a graphic demonstration of nuclear weapons' power and effectiveness when deployed against conventional ground forces.

effect on the workings of their engines or instruments as it was noted. Subsequent to the artillery and aerial bombardment 30 MiG-15bis fighters strafed 'enemy' positions for good measure.

After the impressive display of conventional firepower ended radiation reconnaissance teams moved in. They established that in the immediate vicinity of the point of detonation (epicentre) the level of radiation was 50 roentgens per hour (R/h) one hour after the explosion, while the corresponding levels at a distance of 300m, 500m and 850m were 25R/h, 0.5R/h and 0.1R/h respectively. Next, at noon, the main body of the ground forces went onto the offensive – armour and mechanised formations first, though because of obvious tactical requirements attacking infantry accompanying tanks were dismounted and moved on foot a short distance behind the armour, while second wave infantry formations following on were transported by trucks. It should be noted, that of the many thousands of officers and men taking part in the manoeuvres, about 3,000 directly crossed the zone of the nuclear blast, with all of them having been issued with protective equipment. As to the exercise's overall conduct, rather unsurprisingly the 'enemy' was 'forced into a retreat' and the manoeuvres concluded in a manner which satisfied Soviet military officials observing them.

The immediate aftereffects of the nuclear explosion were subjected to study. The radioactive cloud was propelled by the wind and drifted with a speed of 84km/h at an altitude of 7,000-9,000 metres, and was followed by an appropriately equipped Li-2 aircraft. Of obvious interest were the effects of the nuclear weapon detonation upon vehicles and military equipment positioned in the targeted area. Not surprisingly tanks, in particular when dug in, while not indestructible proved to be most survivable. In contrast, thin-skinned vehicles and aircraft were much more prone to damage. It is interesting to note that of the aircraft placed as targets (22 Il-10 ground attack aircraft, 20 La-15 fighters and two Li-2 transports), five Lavochkins which were over 2,000m from the point of explosion escaped serious damage, as did three Shturmoviks which were 3,500m away.

A still from the film, taken on the ground, showing the 'mushroom' caused by RDS-3, deployed during Exercise Snezhok.

Near the target zone, numerous additional targets were deployed, including captured German guns from the Second World War – like the PAK40 visible here. The piece apparently suffered extensive damage that rendered it non-operational.

Of even greater importance were the nuclear explosion's effects on living organisms. Farm stock suffered death at a distance of 1,200 metres from the point of detonation, serious injury at 1,500m and burns at 4,000m. However, those shielded by woods had a much higher chance of survival, with death occurring at 600m, while serious injury and burns occurred at 1,200m. Similarly, occupying dugouts, trenches and even being on the opposite slope of a hill substantially increased the chance of survival for the animals – thus one can assume for men too. At this point one cannot but pause to ponder about death and suffering. Images of carcases as well as of surviving but burned or injured animals recorded after the explosion inevitably make one think of the terrible suffering men would have to endure had they been subjected to a nuclear explosion.

There were other effects of the nuclear blast as well – both short and long term ones. Fires broke out in the villages Makhovka, Yelshanka-2 Ivanovka, Orlovka and a few others which were located in the vicinity of the exercise area. While the blazes were swiftly put out by the troops, and the people who lost property were compensated, this was

Hard on the heels of the nuclear bomb, air strikes and artillery barrages, dismounted Soviet ground troops moved through the 'combat zone', supported by T-54 tanks – at the time the cutting edge of Soviet armour. Visible in the foreground is another destroyed PAK40.

Soviet infantry attacking. Obviously, the troops were deployed without any kind of protective gear - even without such relatively simple items as gas masks, which were a standard issue at the time. Although not crossing the very centre of the nuclear explosion, they did pass very close by, and were seriously exposed to nuclear radiation.

by far not the end of the story: there was a noticeable increase in the number of cancer patients among the populace living in the vicinity of the nuclear explosion. Unfortunately, exact numbers are hard to come by because the Soviet authorities ensured there were no relevant statistics and records kept. In addition, the troops participating in the manoeuvres were also affected. The protective measures and subsequent decontamination were not always fully implemented or properly carried out. Thus, while numerous officers and men who took part in the war games described lived to see an old age, many also developed various health alignments which could be traced to the effects of radiation exposure during the manoeuvres.

Yet hardly anyone knew about all of this. A few days after the events took place the Soviet daily *Pravda* published in the pages of its 17 September issue a communiqué stating that a nuclear test was conducted. It was asserted that the test's purpose was to examine the effects of a nuclear explosion with the aim of protecting the country from nuclear attack. Afterwards the event disappeared in a proverbial memory hole. All the officers and men participating in the exercise signed a 25-year pledge not to disclose information regarding what they had taken part in. There were hardly any venues to discuss such events in the Soviet Union anyway, but some information began to emerge in the wake of the Chernobyl nuclear power plant disaster, finally leading to its declassification and in December 1991 all the participants of the exercise were given official veteran status. Subsequently a stela was erected in 1994 to commemorate all those who suffered from the radiation aftereffects of the nuclear explosion. Ten years later, in 2004, a tablet was placed in honour of those who participated in the creation of the Soviet Union's nuclear shield.

The Forgotten Bomb

The exercise at Totskoye – while, arguably, best known among such events – was not the only occasion when the Soviets conducted military manoeuvres which involved the detonating of a nuclear weapon. Two years, bar a few days, after the events in Orenburg Oblast, similar war games took place at the Semipalatinsk Test Site. While not a secret for many years, that event is much less well known, not least because while the veil of secrecy has long been lifted, the information available concerning it is much more sparse. Yet despite the fact that the information published concerning those war games is far from exhaustive it is a piece of history which deserves to be related.[2]

Before recounting the events, it is fitting to devote a few words to the idea behind the exercises in question. It is obvious that a nuclear strike has the potential to substantially disrupt enemy defensive efforts. However, in order to fully exploit its results swift action was essential and a heliborne insertion of forces was one of the possibilities for such. The 1950s saw a number of developments including the introduction of the Mi-4 helicopter by the Soviets. This was the first medium lift helicopter in their service, which in turn made heliborne operations a realistic option for the Soviet armed forces. The scenario of the manoeuvres envisioned a nuclear strike against a target behind enemy lines, the effects of which would be exploited by a heliborne assault. In turn, the latter would be followed with a conventional attack on the ground by troops that were to link up with the forces helicoptered-in earlier. The men to be landed by helicopter were drawn from the 2nd Battalion (sans one company) of the 345th Regiment of the Soviet Airborne Forces, augmented by a battery of 57mm guns, a battery of 82mm recoilless guns and a battery of 82mm mortars. The follow-up contingent had a *divizion* (artillery battalion) to provide indirect fire from covered positions, two batteries of guns for direct fire support and a battery of light ASU-57 self-propelled guns (the ASU-57s could not be flown in by the helicopters due to their dimensions and weight). In all, about 1,500 officers and men were involved in the exercise. The assembly point for the forces partaking in the manoeuvres was located some 23km from the mock enemy 'front line' and about 36km from the intended target of the nuclear bomb.

The date chosen to conduct the exercise was 10 September 1956. The necessary arrangements were made beforehand, including the preparation of 'enemy' positions with trenches dug and practice targets set up. On the chosen day, a Tu-16 delivered the nuclear weapon dropping it from an altitude of 8,000 metres. The nuclear bomb of 38kt yield detonated at an altitude of 270 metres with the time recorded being 09.00 and exploded at the coordinates 50.4°N and 77.8°E, some

An Image Worth a Million Words

A picture is worth a thousand words the old adage goes. It is indeed so, those ideas which are complex or otherwise hard to express in a succinct manner can be conveyed by a single image which puts across their meaning or essence more effectively than even a lengthy written or verbal description.

This very much applies to the prospect of a nuclear war, which unfortunately still remains a possibility. It would be nothing but a catastrophe for humanity, the shadow of which since the end of the Second World War has hung over generations past, and still looms over the present as well as future generations. Few images convey this as clearly as the photograph taken on a sunny late summer morning by Ivan Sharonin. Residing in the village Sorochinsk he worked as the musical director at the local *dom kultury* – literally a house of culture – an institution peculiar to the former Soviet Union and other Eastern Bloc countries which was meant to serve the cultural and artistic needs of a given community. On 14 September 1954 he was out on a village street when he saw a nuclear explosion in the distance. Wasting no time Sharonin reached for his camera and took a picture. However, being in awe of the unfolding events as well as in a hurry to capture the moment on camera, he forgot to wind on to the next frame before taking it. As a result, the image of the ominous mushroom cloud formed in the course of the Totskoye exercise was superimposed onto a photograph of children which he happened to have taken beforehand.

Despite 66 years having passed since the image was taken, its powerful message still holds. While worldwide pandemics and other calamities mankind currently faces tend to capture our attention, it should be remembered that we are not out from under the mushroom cloud's shadow yet. A nuclear apocalypse was menacing men, women and children decades ago, and continuous until this very day. Even if we tend not to think about it on a daily basis, this does not make this dreadful threat any less real. More could be said and written in this regard, but one glance at Ivan Sharonin's photograph should suffice for it is worth a thousand – if not a million – words.

A pre-mission rehearsal for officers involved in the exercise of 10 September 1956. Models were used to better visualise the movement of dozens of helicopters and other involved aircraft.

80 metres from the intended aiming point. About 25 minutes after the bomb went off, radiological reconnaissance teams entered the zone of the explosion and it was decided to land the troops circa 650 – 1,000 metres from the epicentre of the detonation. Within 43 minutes of the explosion the troops started to land – in all 272 troops with additional heavy weapons were brought in by 27 Mi-4 helicopters. After being air-landed it took the troops about 17 minutes to take up defensive positions and prepare to fend off enemy counterattacks so that they could hold out until the main force linked up with them in accordance with the war game's scenario. Apparently, all went according to plan and within two hours of the time of the nuclear explosion the exercise was concluded.

The events described were known in the West as Joe-26. It was the second and last Soviet tactical exercise involving the airdrop of a live nuclear weapon followed by an actual deployment of troops.

The live firing exercise of 10 September 1956 – undertaken at the Semipalatinsk polygon and known as 'Joe 26' in the USA – saw the involvement of an entire regiment of Soviet airborne troops and 27 Mil Mi-4 helicopters.

Officers of the 345th Airborne Infantry Regiment embarking on one of the Mil Mi-4 helicopters involved in the exercise of 10 September 1956.

A formation of Mi-4s in the process of taking off for their fateful participation: capable of carrying a vehicle like a GAZ jeep, or a 57mm gun, the Mi-4 provided useful medium lift capability.

A pair of Mi-4 with troops on board underway into the 'combat zone'.

Airborne troops of the 345th Regiment seen inside one of the Mi-4s. Notably, this time they have received at least gas masks and protective boots and appear to be wearing oversuits.

A radiological reconnaissance team in full protective gear. The man in the foreground appears to be checking a radiation dosimeter.

Close-up of a Soviet radiation dosimeter. The particular device pictured shows a reading of just above 0.04R/h.

A Tu-16A of the 35th Independent Test Aviation Regiment – the core element of the 71st Special Polygon – approaching the target zone for its drop.

A Soviet bomber pilot with personal protective equipment which constituted breathing apparatus with a filter as well as protective goggles

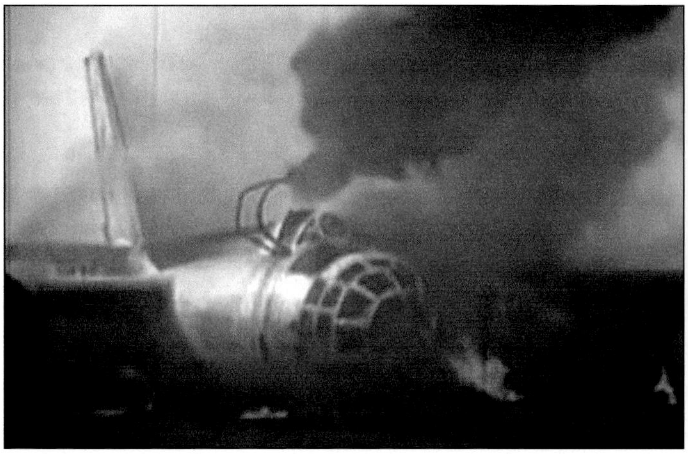

Ironically, one of targets 'spent' during the Snezhok exercise was this disused Il-28 bomber: made of aluminium, the skin of the aircraft quickly caught fire when hit by the heat wave released by the RDS-3 nuclear bomb.

A Path to Certain Victory?

Having described Soviet manoeuvres which involved the detonation of a nuclear weapon, the author cannot help but to comment. First of all, it is obvious that 'playing nuclear war' was a very costly kind of entertainment. The human part of the bill was footed by the military men directly involved and civilians alike, in addition to the environmental damage that cannot be discounted either. It is a fact that the Soviets were not the only ones engaging in such testing. The US military conducted a series of nuclear explosions in conjunction with observer programs, tactical manoeuvres and damage effects tests collectively known under the codename Desert Rock. However American exercises involving nuclear weapons tests are not a subject matter of this volume.

Having said all this, the most important question arising is: of what value might have this been in case of an actual war? Understandably it may be very tempting to cut a Gordian Knot of tactical problems with a nuclear weapon. It is not hard to imagine circumstances under which the temptation may even turn into an irresistible urge. There is no doubt that should the Soviets have attacked, a defending Western force could be obliterated with a nuclear weapon or weapons. Yet by the same token, Soviet forces advancing to exploit the resulting situation could be swept away with American (British or French) nuclear weapons like dry leaves. The latter was likely in so far as the use of a nuclear weapon by one side would have removed any inhibitions to nuclear use felt by the other side. As a result, the use of nuclear weapons might have made victory unattainable, or in the best-case scenario resulted in a pyrrhic victory – at the cost of countless men killed, maimed in horrendous ways and poisoned by radiation (not to mention material losses) the 'victor' would end up as a proud possessor of a radioactive wasteland.

3
THE MESSENGER OF DEATH

While Soviet military leadership was preparing to 'play' nuclear war, Soviet scientists were about to deliver new 'toys': even more powerful weapons. They had good reason to work on these new weapons as the United States tested its first thermonuclear device on 1 November 1952. However, their counterparts from across the Iron Curtain were not far behind; on the contrary, they were well advanced in building a weapon of this type on their own. To this end crucial theoretical ideas as well as practical research were contributed by the nuclear physicist and weapons designer Andrei Dmitrievich Sakharov.[1]

A casing of the RDS-6 on display at the Nuclear Weapons Museum in Sarov.

Getting There, but Not Quite

Soon enough the Soviets had their nuclear device of a new type and exploded it on 12 August 1953 at Semipalatinsk – 50.2548°N and 77.4948°E to be exact – the event becoming known as the Joe-4 test in the West. The RDS-6, as it was designated by the Soviets, was placed atop a tower. For the purpose of the test buildings were constructed and military equipment, such as tanks and aircraft, positioned at varying distances from the test rig. The explosion, which had a yield of 400kt, predictably devastated everything in the vicinity. Its power was illustrated by the fact that tanks which weighted many tons were flung on their sides or in other awkward position as if they were mere toys tossed by an ill-behaved child.

Despite the test being described as successful, the Soviets could not yet rejoice at having fully caught up with the Americans because their thermonuclear device was built as a layer design where the primary is wrapped in shells of lithium and tamper – the Soviets even reflected this by unofficially calling it a *sloika*, which is a layer cake in Russian. The design resulted in a workable thermonuclear weapon but one that could not be scaled up to megaton yields. This was no secret in the West, being concluded from available information by the Bethe panel (named so after Hans Bethe, a German-American nuclear physicist).[2] However the Soviets were far from being finished.

The 'Real' Hydrogen Bomb

Not satisfied with the results achieved thus far, the Soviets, with Andrei Sakharov at the fore of the effort, developed a two-stage fusion-type thermonuclear device; that is a 'real' hydrogen bomb. Importantly the RDS-37, as it was designated, was right from the start built not as an experimental device but to be a deployable weapon.

The first drop of the Soviet RDS-37 thermonuclear bomb was preceded by considerable preparatory work and tests. Obviously, the aircraft – a Tu-16 jet bomber – which was to perform the mission had to be fitted with appropriate on-board systems to carry and drop the bomb, as well as protective modifications, such as a white anti-flash finish, the sealing of hatches and covers, and covering some of the more fragile pieces of equipment, wiring and fittings with insulating materials – altogether about 30 modifications in preparation for the mission were made. In addition, before a live bomb was air dropped and detonated, the decision was made to first test it's in-flight ballistic characteristics. For this purpose, two mock-ups fitted with telemetric equipment were dropped: the first on 17 November 1955 from an aircraft flown by W. F. Martynenko and a second one two days later, the mission being performed by F. P. Golovashko. Having done all this the date of the actual live test was set for 20 November. Since all

A Tu-16A of the 35th Independent Test Aviation Regiment about to be armed with a thermonuclear bomb. Its weapon bay is open with a 'curtain' which was both to provide protection from the elements as well as from unwarranted view while the bomb was loaded.

The RDS-37 thermonuclear bomb being wheeled to the bomber.

A front view at the RDS-37, on its way into the bomb bay of the Tu-16.

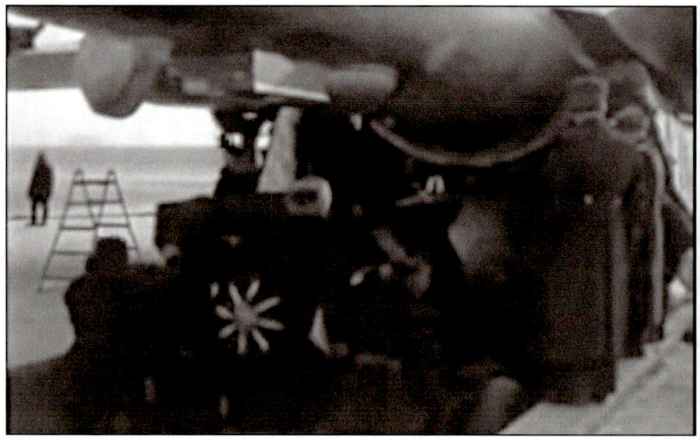

Rear view at the RDS-37 and the underside of the Tu-16, as the bomb was being hoisted into the bomb bay. Because of the size of the RDS-37, the aircraft had to be positioned over a pre-prepared 'loading trench' into which the bomb was lowered and from which it was then hoisted into the bomb bay. Of note is the breaking parachute pack in the rear of the bomb.

About to 'disappear behind the curtain', the last view of the RDS-37 – the first Soviet hydrogen bomb. Exploding a device of this type gave the Soviets new capabilities and demonstrated that they were at the forefront of nuclear weapons development.

was readied and tested everything should go according to plan, or so it seemed.

If Anything Can Go Wrong …

The process of arming the bomber started in the morning hours of 20 November, at 06.45 to be precise, with four hoists lifting the RDS-37 into the Tupolev's bomb bay. This, as well as other procedures, took less than two hours to complete. At 09.30 the bomber took off from Zhana Semey with Major Golovashko at the controls setting course for the Semipalatinsk Test Site.

Weather forecasts spoke of good meteorological conditions: alas they unexpectedly began to deteriorate. A cloud layer above the target area obscured it, making visual aiming impossible, while a technical malfunction rendered the Rubidiy bombing radar inoperable, with all attempts to repair it being unsuccessful. Thus, the aircraft had on-board a live thermonuclear weapon ready to be airdropped, yet this was impossible because the target could not be visually sighted and a problem with the radar prevented the use of available technical means to do so.

Obviously, a solution to this precarious situation had to be found – and in short order – because the Tupolev's fuel supply limited the time it could stay airborne. The idea of dropping the bomb over some remote area was contemplated but rejected because it could not be guaranteed that the blast would not destroy one or more settlement: the vast expanses of Kazakhstan while thinly populated was not totally uninhabited. Another problem would have been radioactive contamination outside the designated testing grounds. Considering these points, it was decided that the Tupolev should return to base and land with the bomb. This involved a number of risks, in particular it was feared the bomb might explode. While nuclear bombs have a number of safety mechanisms to prevent their accidental detonation, they also work the other way around; that is they are also built so as to ensure they will explode once dropped. It was precisely the latter which was, despite assurances of the experts present, causing much anxiety as the bomb was armed and ready to be released. In particular, it was feared that the barometric sensor might set the bomb off once the aircraft descended to the altitude at which it was meant to detonate. Fortunately, and to everybody's great relief, nothing of the sort happened.

The landing itself was not without problems, either: a sandstorm enveloped the airbase and the pilot missed the runway on his first

The Tu-16A carrying RDS-37, in the process of taking off. Notably, by the time of this test – in November 1955 – the aircraft had had its undersides painted white for improved protection from the flash and heat created by a thermonuclear detonation.

The Tupolev Tu-4 bomber was a Soviet clone of the American B-29, as is clear from even a cursory look. Yet it was not an exact copy: it differed from the US aircraft in the thickness of its skinning, its engines and the calibre of its defensive armament. Although quickly becoming obsolete, the Tu-4 remained highly significant for it plugged a gap in Soviet aviation technology development. Most importantly, it also gave the Soviets their first nuclear capable bomber: a capability which was not just theoretic, as demonstrated when on 14 September 1954 the Tu-4 illustrated here – construction number 2502287 – dropped a live nuclear bomb in the course of the Totskoye military manoeuvres. (Artwork by Tom Cooper)

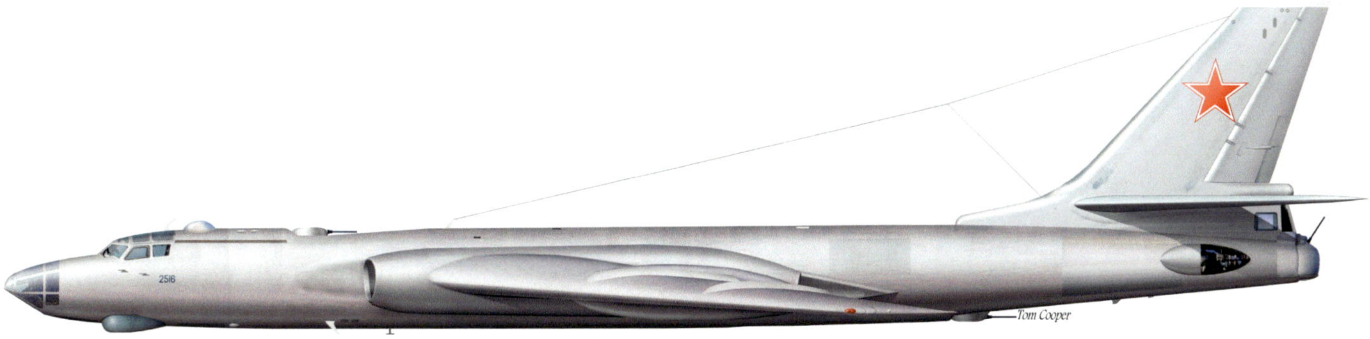

Of all the Soviet medium and heavy turboprop and jet bomber types, the Tupolev Tu-16 was the most numerous. It was heavily involved in the live testing of Soviet nuclear and thermonuclear bombs. The aircraft pictured here dropped the first Soviet thermonuclear bomb on 22 November 1955. Because of the bomber's whiteish anti-flash finish, the Soviet nuclear physicist and weapons designer Andrei Sakharov associated it with the 'messenger of death'. Witnessing this man-made doomsday machinery in action was among the factors which led Sakharov to have second thoughts about his work and the system his toils served, and in due course to he became one of the best known Soviet dissidents. (Artwork by Tom Cooper)

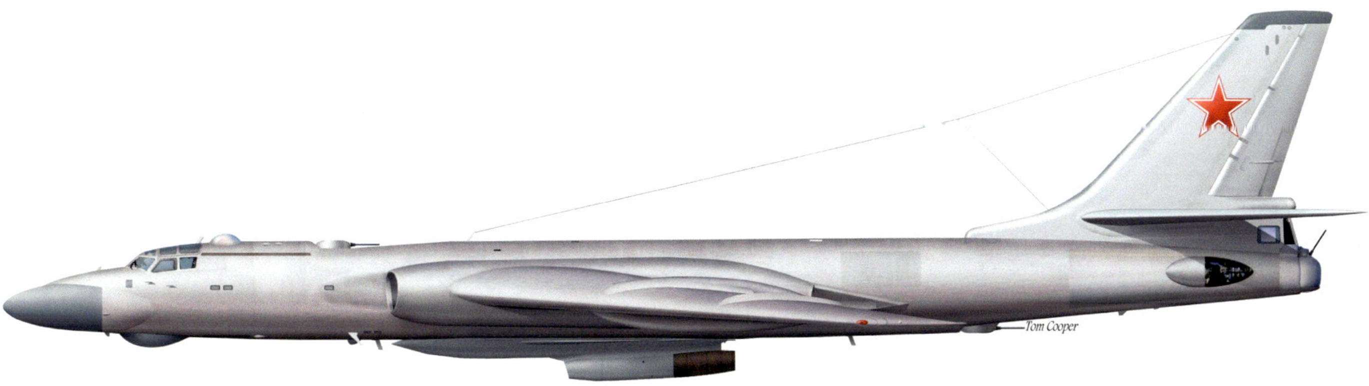

The Tu-16 proved a versatile design, adaptable in a number of different roles. One of the most prominent was that of a *raketonosets* – the missile carrier – which resulted in the bomber serving as a launch platform for early land-attack cruise missiles and anti-ship missiles, all of which could be armed with nuclear warheads. The Tu-16K-10 was one of the most powerful sub-variants of this kind: it could be armed with the K-10 missile – the installation of which required the replacement of the glassed nose covering the navigator-bombardier's position, with a large housing for a guidance radar. The huge weapon did not fit entirely into the bomb bay, but was installed semi-recessed, with the entire lower half of the fuselage and engine nacelle protruding underneath the aircraft. At least two K-10s equipped with live nuclear warheads were test-fired in the early 1960s. (Artwork by Tom Cooper)

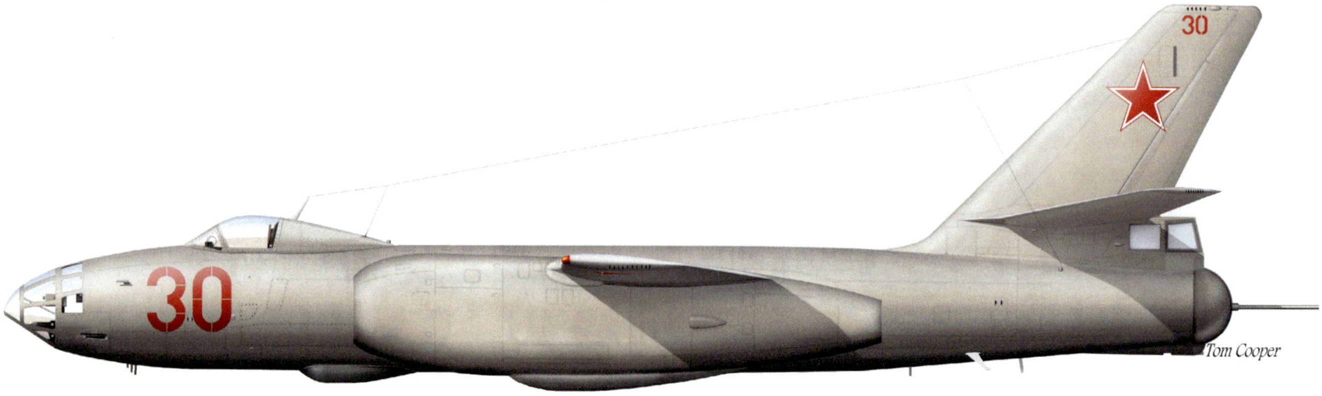

Ilyushin Il-28 light jet bombers (though sometimes referred to as a medium) formed the backbone of Soviet tactical offensive aerial capability. Once the RDS-4 nuclear bomb was developed, its compactness enabled the Il-28 to carry out tactical nuclear strikes. Such a weapon was live tested from the Il-28 illustrated here on 23 August 1954. It was only the second Soviet live test of an air dropped nuclear bomb, but also the first of a compact tactical nuclear weapon, and the first from a jet powered aircraft. The last two factors made this test particularly important in the Soviet nuclear weapons program. (Artwork by Tom Cooper)

The Mil Mi-4 was the first Soviet medium lift helicopter. Once it entered service in substantial numbers it gave the Soviets the ability to mount large-scale operations such as inserting troops behind enemy lines, or rapidly exploit the results of a nuclear strike by means of a heliborne landing. While the Mi-4 could not transport even light armour such as the ASU-57 self-propelled guns, it could carry jeep-sized vehicles, anti-tank guns and mortars, meaning the troops it deployed could receive light fire support. The helicopter's ability to carry personnel, equipment and vehicles was also very useful when performing auxiliary tasks in support of nuclear testing. (Artwork by Tom Cooper)

The Mikoyan i Gurevich MiG-15 jet fighter is almost synonymous for the first decade of the Cold War – foremost for the impact it had during the Korean War and the impression it left in the West. The fact that it could outmatch its Western competition at the maximum operational altitude came as a rude surprise, especially for the USA. Although manoeuvrable and packing considerable punch – thanks to its heavy cannon armament, including two 23mm and one 37mm gun – it could not carry nuclear ordnance. It did take part in related testing however, and a unit of these early MiG fighters – including the example illustrated here – took part in the Totskoye exercise that involved a live air drop and detonation of a nuclear bomb. (Artwork by Tom Cooper)

The 'Tsar Bomba' had such dimensions that it took a specially modified Tupolev Tu-95V bomber to carry, as was true for its derivatives with lesser yield which used the same bomb casing. Large by any standard, the size of the bomb casing is best illustrated by the fact that it could not fit even into the huge bomb bay of the Tu-95: it still protruded underneath it. While the bomb was immensely powerful its practical value was limited because its sheer size made it imperative to enlarge the bomb bay at the expence of fuel tanks, reducing the fuel supply, which would be necessary for the aircraft to reach intended targets in the continental USA. Notably, the Tu-95V that deployed the Tsar Bomba had most of its lower fuselage and undersides of its wings and fin spray-painted in anti-flash white. (Artwork by Tom Cooper)

The Myasischev M-4 – and its improved version, the 3M, illustrated here – were no less ungainly than their US counterpart, the Boeing B-52, but definitely less successful in their intended role. Despite numerous improvements vis-à-vis the M-4, including an expanded fuel capacity and more fuel efficient engines, the 3M's unrefuelled range of 11,850km was barely sufficient to reach the USA and return to the USSR. With the help of in-flight refuelling, the type proved capable of flying 15,400km, but in turn its VD-7 engines required very frequent maintenance inspections and ultimately had to be replaced: their replacements, VD-7Bs, had been developed by 1960, but proved less fuel efficient, reducing the range to 9,400 kilometres. Because the type also proved difficult to maintain and fly it served only relatively briefly as a strategic bomber and was soon after modified for many other tasks, including reconnaissance and as a tanker for in-flight refuelling. (Artwork by Tom Cooper)

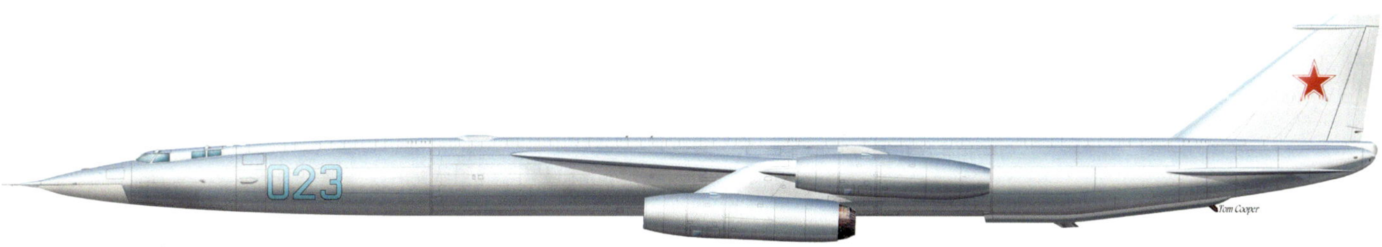

Its slim and elongated fuselage, and a huge delta wing made the Myasischev M-50 bomber appear like an oversized fighter aircraft, if not from a science-fiction movie. The most unusual solution was the installation of the outer engines in wingtip pods. While never put into production, the M-50 was still used to confuse Western intelligence: and unsurprisingly, the prototype received several different bort numbers. At the Tushino Air Show on 9 July 1961, it wore the bort 'blue 12', but later on this was repainted into 'blue 023' (outlined in black). In fact, wowing spectators at that air show was about the most this aircraft has ever achieved: ultimately, it proved to be another of the Soviet unfulfilled hopes. Beyond that it actually flew, the only other satisfaction for its designer was that it was foremost the insufficient performance of the available engines – rather than any kind of design-flaws – that ended its career before it even began. (Artwork by Tom Cooper)

To withstand freezingly low temperatures encountered during operations at high altitudes, Soviet Tu-4 crews received heavily padded jumpsuits, thick leather gloves and overboots made from animal skin. Their earliest flight helmets were made of leather and, apparently, derived from the M-43 model that was in widespread use during the Second World War. The parachute harness was of the type introduced in 1947: thanks to a continuous upgrade of details, this remained in service until the late 1970s. (Artwork by Anderson Subtil)

Service uniforms of Soviet Il-28 crews in the 1950s were relatively simple. They consisted of ShL-50 leather helmets with microphone and earphones, ZS-50-5 flight jackets, standard fabric shirts, ties, and trousers (often enhanced through a pullover), and yuft (a kind of thick, soft leather) jack boots. The board – usually attached to the left wrist – was used to carry maps and other operations-related documents. (Artwork by Anderson Subtil)

The first 'Combined Arms Chemical Protective Suit' for the Soviet infantry was designated OZK OP-1. Probably introduced to service in the late 1950s, it was manufactured by Slavyanskaya Clothing Factory, and consisted of three main components: the OP-1 protective coat, a pair of boots and gloves, all manufactured of rubberised, waterproof fabric. In this illustration, the OP-1 jacket is shown worn as an overcoat: alternatively, it could be worn as a coverall, with the lower parts of the suit attached around the legs. Under the suit, the soldier still wore the SSh-40 or SSh-60 steel helmet, and the ShM-1 gas mask. This infantryman is shown armed with the initial version of the Kalashnikov AK-47 assault rifle. (Artwork by Anderson Subtil)

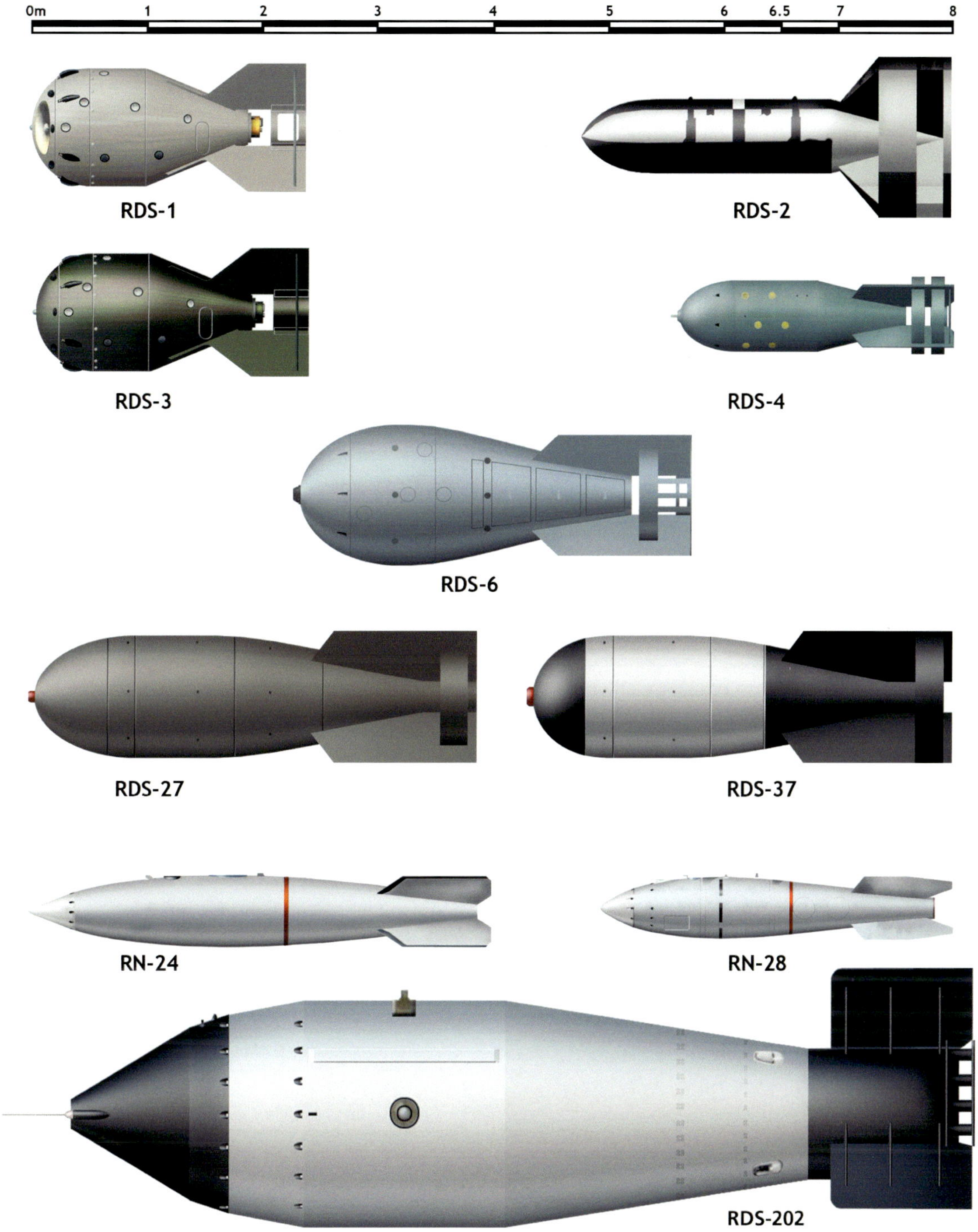

In the course of the Cold War the Soviet Union became a nuclear superpower, eventually possessing a large arsenal of nuclear weapons varying in size, yield and type.

The RDS-1 – an implosion lens design utilising plutonium and based on the 'Fat Man' design of the US origin, obtained through espionage – was the first Soviet nuclear bomb. It was a bulky weapon weighing around 4,600kg. Notable were its prominent radar fuses at the front, giving it the appearance of 'eyes'. The RDS-2 was an improved version of the RDS-1, but with a new core and new explosive lenses, providing it with a yield of 38kt. The weapon weighed around 3,200kg. While neither RDS-1 nor RDS-2 seem to have ever been air dropped, the RDS-3 was: this weapon was of similar size and shape to the RDS-1, but had a composite design including a 25 percent plutonium core inside a 75 percent uranium shell (the Soviets were short on plutonium, but had 'more than enough' uranium). Despite some misgivings, it was successfully tested with a maximum yield of up to 42kt.

Colloquially known as 'Natasha', weighting about 1,200kg, and being notably smaller than earlier weapons, the RDS-4 was the first Soviet 'tactical nuclear bomb'. Nevertheless, it still demonstrated a yield of 28kt during live testing. Its significance lay in the compact design, enabling carriage by aircraft lighter than the Tu-4.

The RDS-6 was the first Soviet thermonuclear bomb, though with some caveats, including its layer design. While this proved a workable solution, the weapon's yield could not be scaled up, and the design was bulky, weighing around 5,000kg. Therefore, the first 'real' Soviet thermonuclear bomb became the two-staged RDS-27, followed by the RDS-37, each weighing around 5,570kg, which had a yield of up to 3Mt (though mostly scaled-down to around 1.6Mt during testing). Eventually, during the 1960s the Soviets managed to downscale their tactical nuclear bombs, such as the RN-24 and RN-28, to a size that made them compatible with fighter-bombers. Of course, all these weapons were dwarfed by the RDS-202: the largest man-made explosive device ever, over 8 metres long and weighing over 27,000kg. (Artworks by Tom Cooper)

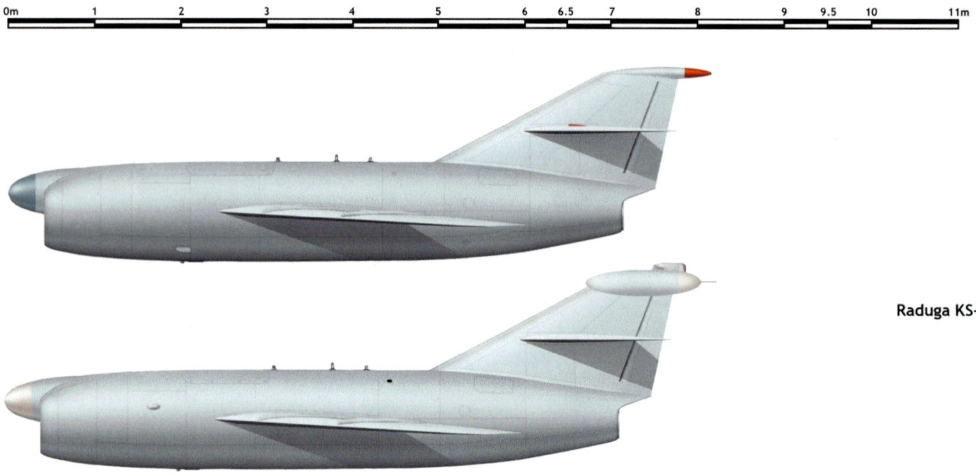

Raduga KS-1 (ASCC/NATO-codename 'AS-1 Kennel')

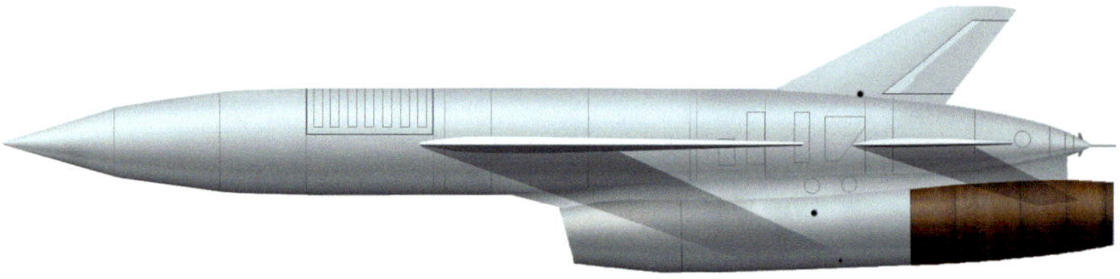

Raduga K-10S (ASCC/NATO-codename 'AS-2 Kipper')

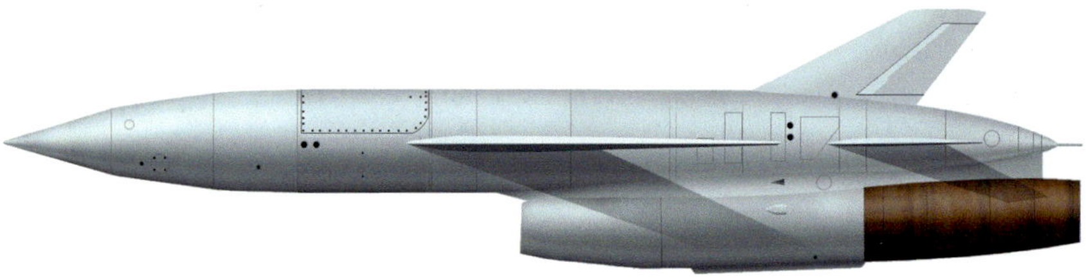

Raduga K-10SM (ASCC/NATO-codename 'AS-2 Kipper')

Raduga Kh-20 (ASCC/NATO-codename 'AS-3 Kangaroo')

Raduga KSR-2 (ASCC/NATO-codename 'SA-5 Kelt')

Through the 1950s, and especially in the 1960s, a series of cruise missiles designed by the Raduga OKB became the primary armament of the Soviet bombers. The first to enter service in substantial numbers was the KS-1 Kometa (Comet), codenamed 'AS-1 Kennel' by the ASCC/NATO. It was fielded in several variants including land-attack and anti-ship versions (top profile). Essentially, the missile was an unmanned jet aircraft meant to perform the mission of delivering a conventional or nuclear warhead. The latter was of crucial importance for at least partially mitigating Western – and particularly US – naval supremacy: a single KS-1 armed with a nuclear warhead was capable of obliterating any vessel. The dielectric nose housed an active radar homing system, while the blister on the air-launched version (shown second from the top) contained a rearward facing antenna used to communicate with the launch aircraft. Although quickly falling into obsolescence, the KS-1 proved an important step in further development of the Soviet weapon systems of this type. The air-launched variant was deployed by Tu-4 and Tu-16s of the DA, and exported to Egypt and Indonesia.

Like the mass of Soviet military aircraft of the 1950s and the 1960s, Soviet cruise missiles were sparsely marked and usually painted in a greyish light-blue or bare metal overall, which gave them an anonymous look – intentionally. The K-10S (third from the top; ASCC/NATO-codename 'AS-2 Kipper') was no exception to this rule: this large, air-launched anti-ship missile was a respectable conventional weapon system at the time it entered service, further augmented by its ability to carry a nuclear warhead. The K-10SM (fourth from the top) was a low altitude variant, meant to present a most difficult target for all types of contemporary air defences, forcing potential opposition to search for ways to shoot down its carrier before the missile was released.

Developed from the design of the MiG-19 interceptor, the turbojet powered Raduga Kh-20 (ASCC/NATO-codename AS-3 Kangaroo) was the biggest of the Soviet cruise missiles of the 1960s.

The KSR-2 (ASCC/NATO-codename 'AS-5 Kelt') was the final air-launched land-attack and anti-ship missile to enter service with the DA during the period covered in this book. It represented a substantial improvement in virtually all characteristics in comparison to its predecessor. Instead of an air-breathing engine it was powered by a two-chambered rocket motor. Foremost deployed by Tu-16s, and meant to be used to deliver conventional warheads, it could alternatively be fitted with a nuclear device. The KSR-2 and its anti-radar variant - the KSR-11 - was exported to Egypt and successfully deployed in combat during the Arab-Israeli War of October 1973. (Artworks by Tom Cooper)

Bottom view of a Tu-4 – the first Soviet strategic bomber, and the type that enabled them to develop at least a limited nuclear strike capability in the early 1950s – showing the doors of its front bomb bay in the open position. (Albert Grandolini Collection)

A still from a documentary film, showing an RDS-6 nuclear bomb shortly after release from a Tu-4 bomber, during one of the live drops in the 1950s. (Albert Grandolini Collection)

An RDS-6 bomb in the process of being re-loaded from a train to the Tu-4 bomber that was to deploy it during a test. (Albert Grandolini Collection)

One of the best stills showing the specially modified Tu-95V piloted by Lieutenant-Colonel Durnovtsev seconds after release of the RDS-202 on 30 October 1961. (via Tom Cooper)

A view of the centre section of the Tu-95V, revealing details of the installation of the RDS-202, the Tsar Bomba. (Albert Grandolini Collection)

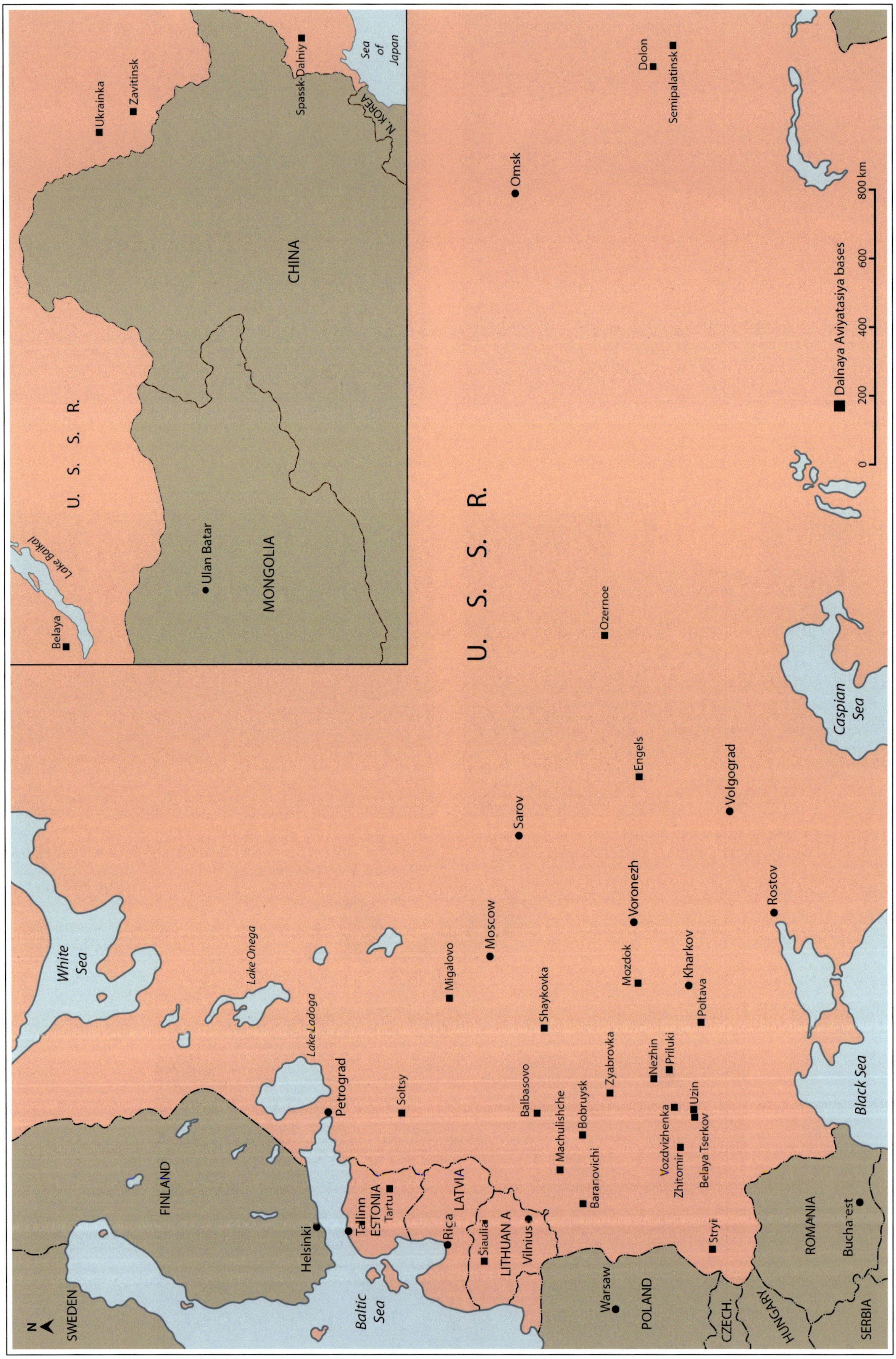

All the known air bases of the DA in the USSR as of the 1950s and the 1960s. (Map by George Anderson)

A good view of the RDS-37 hurtling down in free fall before its breaking chute opened.

approach. The second attempt was better and the aircraft touched down remarkably smoothly. This also did away with another potential cause for disaster as the shock of the undercarriage coming down hard on the runway could have resulted in the bomb breaking loose from its attachment points in the weapons bay. Fortunately, all went well, with the time of the landing being recorded as 12.00 (which meant the mission had taken two and a half hours). The bomb itself was carefully removed between 14.15 and 16.00, subsequently being taken in hand to prepare it for another test attempt. After all efforts, this test had been a failure, one due to such capricious factors as the weather and an unexpected technical malfunction.

Success, at Last

Preparations for another test attempt were made without delay. Thus, only two days passed from the first try when on 22 November 1955 the readied bomb was brought to the airfield's tarmac at 04.50. At

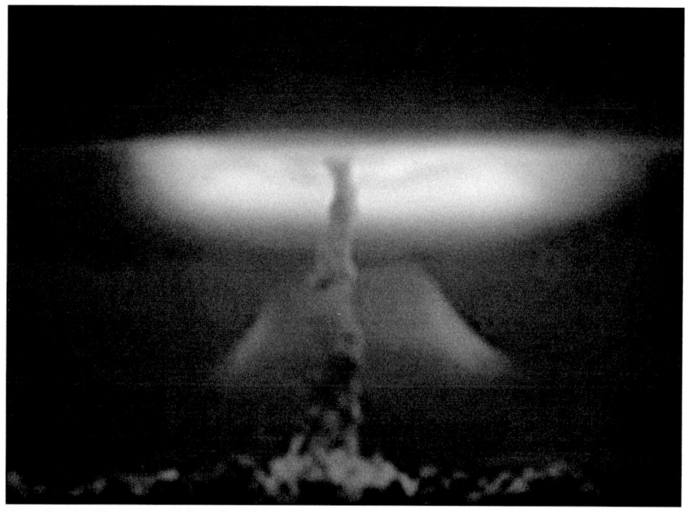

The fireball created by the detonation of the first Soviet thermonuclear bomb.

06.55 the RDS-37 was lifted into the Tupolev's weapons bay for a second time and it was 08.34 when the aircraft took to the air. Its crew comprised of:

- Major Fedor Pavlovich Golovashko: senior test pilot
- Captain I. M. Romenskiy: the second pilot
- Major A. N. Kirilenko: navigator
- Captain V. I. Lazarev: navigator-radar operator
- Lieutenant B. I. Ozherelyev: radio operator
- Starshina (an approximate equivalent to sergeant-major) N. P. Suslov: defensive armament operator/gunner

In addition to the Tu-16 carrying the bomb, a number of other aircraft were involved in the test: an Il-28 was tasked with observing the explosion's aftermath as well as collecting atmospheric samples, subsequently radiation levels were also monitored by appropriately equipped Yak-12 and Li-2 aircraft. Apart from the aircraft which were to carry out the test and monitor its aftermath, pairs of MiG-17 fighters were tasked with securing the airspace.

Returning to the test itself, the Tu-16 flew at an altitude of 12,000 metres with a speed of 870km/h (some sources give the speed as 985km/h), realising the bomb over the intended target at 09.47. Soon thereafter its 6m^2 parachute deployed and the bomb descended to 1,550m, at which altitude it detonated, the coordinates being given as 50.42°N and 77.78°E. Meanwhile the aircraft was about 15km from the point of detonation when the bomb exploded. First its crew experienced dazzling light, considerably brighter than a sun at noon, and then, 224 seconds after the explosion, the shockwave caught up with the bomber – as a result the Tupolev was lifted up and accelerated by 2.5g. Fortunately the aircraft escaped any serious damage and returned to base without incident.

The explosion manifested itself by a bright flash lasting some 10-12 seconds, and circa 5 to 7 minutes after the explosion the resulting mushroom cloud reach a diameter of about 25-30km and an estimated height of 13-14km. While the bomb should have had a yield of approximately 3Mt, it was intentionally reduced to about 1.6Mt (though estimates of a somewhat higher actual yield are also given) by having the fusion cell modified through the replacement of some of the fusion fuel with passive material. However even after this the weapon still yielded previously almost unimaginable destructive power.

Those gathered at an observation point located 32km from the centre of the detonation experienced first-hand the power which was thus unleashed. First, they were exposed to intensive heat and

The Tu-16 bomber with its deadly war load of a thermonuclear bomb. This is the view as Andrey Sakharov saw it: no wonder to him it became a sinister bird of prey, a white messenger of death.

Major Golovashko, pilot of the Tu-16A that deployed RDS-37 on 22 November 1955.

subsequently heard and saw the shockwave approaching. The only way to avoid injury was to lay face down with feet pointed in the direction of the blast. Within 90 seconds the shockwave reached and passed over them, after which they leapt up in jubilation as the test was an obvious success. Yet not everyone was in such a buoyant mood as will be described below. The power of the blast was also clearly felt at the command post from which the test was directed where, despite it being about 70km from the point of the explosion, some people were knocked down. It should be added that accounts of how the explosion was experienced at the observation point and the command post differ in some details, depending upon the source, though they all confirm that the power of the weapon was clearly felt at both.

A few people being shaken would have hardly been a high price for a successful test of this kind. Alas it also resulted in widespread damage. This is illustrated best by the fact that windows were reportedly shattered at distance of over 200km from the point of detonation. Unfortunately, there were also a number of victims: a trench collapsed on top of soldiers guarding the test site 36km from the explosion; with one of them succumbing to suffocation. A second mortal casualty of the test was suffered in the hamlet Malyye Akzhary where a three-year-old child was killed. In addition, a number of residents of Semipalatinsk City, the village of Mayskoye, as well as the state-owned farms of Voroshilovgradskiy, Stalin-Tuy and Semiyarskoye sustained injuries of varying severity. The weapon was obviously very powerful, but one additional factor which amplified its destructive effects was that the bomb detonated under a temperature inversion layer which reflected a large proportion of the explosive force back to the ground.

The Aftermath

The tests provided other lessons as well. For one, the necessity to work out and implement procedures to safely land an aircraft with a nuclear weapon in its bomb bay became apparent. In addition, the casualties and material damage resulting from this test demonstrated that despite their vastness the expanses of Kazakhstan did not provide enough space for live trials of nuclear weapons with a large yield. Another location had to be sought for this purpose and indeed a remote place suitable to conduct such tests was found as will be described separately.

Even though it was a peacetime test, the Soviets still considered the airdropping of the hydrogen bomb a major feat. Thus, the pilot, Major Golovashko, was on the 11 September 1956 awarded the title Hero of the Soviet Union and received the Gold Star medal (which was a small star indeed made of gold) no 10,846.[3] All of the other crewmembers were also awarded medals.

With this test the Soviets also clearly showed, or rather re-confirmed, that they were to be taken seriously when it came to nuclear weapons development. Namely theirs was the first air-deliverable thermonuclear weapon tested by being dropped from a bomber. The United States' first airdrop of a thermonuclear weapon (Redwing Cherokee) took place about half a year later, on 20 May 1956, while Great Britain's Grapple I/Short Granite test occurred only on 15 May 1957.

There was one more aftereffect of the RDS-37 test. Soviet nuclear physicist and weapons designer Andrei Sakharov started to have second thoughts about his work and the system his toils served. He had witnessed the explosion as well as the bomber, which to him looked like a sinister bird of prey, a white-painted messenger of death, for that colour is associated with death and mourning in some cultures, particularly Asian ones.[4] As a result, over time, Sakharov became one of the best known Soviet dissidents.

4
THE MONSTER BOMB

By the mid-1950s, the Soviets had developed a number of nuclear as well as thermonuclear bombs. Arguably any such weapon is monstrous by its very nature, however, for some it is particularly apt. In the case of the AN602 this term is especially fitting, for this thermonuclear weapon was indeed monstrous both in terms of its yield as well as weight and dimensions, so much so that it deserves to be labelled not only as *a* monster bomb but as *the* monster bomb. This particular bomb went under several designations including RDS-202, RN202 and names such as Ivan. It was also called *Kuzkina mat* literally meaning the mother of Kuzma.[1] Arguably, the weapon remains best known as the Tsar Bomba – even if the Soviets never actually used this term. Since it is pointless to 'swim against the tide', Tsar Bomba will henceforth be used herein as well. The 'doomsday device' was developed by a team of physicists headed by Yulii Borisovich Khariton, which also included many other distinguished Soviet scientists, with Andrei Sakharov among them.[2]

A Special Aircraft for a Special Mission

While considerable attention is usually devoted to the bomb itself and its test, the aircraft which delivered it and the men who flew it usually stay in relative obscurity. The bomb was to be dropped by a Tu-95V, an aircraft devised specifically to carry and drop the Tsar Bomba. In the Russian language the designation is written as Ту-95В which leaves one with the not so obvious question of how to transliterate the Cyrillic character В into the Latin alphabet. In the English language V seems to be the most obvious choice, yet in many languages including the author's native Polish, the letter W would be more appropriate. However, since this is being written in English it shall be Tu-95V.

Officially the task of creating a delivery aircraft for the new bomb was given to the Tupolev design bureau by the Soviet Council of Ministers on 17 March 1956, though work on the project had already commenced before that. Two options were considered; the Tu-16 jet bomber and the Tu-95 turboprop. It swiftly became obvious that the latter was suited best for the task at hand.

Early calculations gave the weight of the bomb as 36,298kg. This was unacceptable as the aircraft would hardly be able to get into the air with such a load and any prolonged flight was out of the question. Thus, a revised designed of the bomb with a calculated weight of 'only' 18,149kg was prepared which showed itself to be more feasible for practical purposes.

Having gotten a realistic starting point for the design of appropriate aircraft modifications work could commence. The project was headed by the Tupolev Design Bureau's Deputy Chief for Armament Systems A. V. Nadashkevich. A standard production Tu-95 aircraft number 5800302 (in some sources just referred to as 302) manufactured in 1955 was taken in. The most obvious changes which had to be made were the fitting of a bomb carrying and release mechanism capable of handling the mass of the Tsar Bomba, as well as the enlargement of the bomb bay. The unusual ordnance would not only be much heavier but also considerably larger than any piece previously carried.

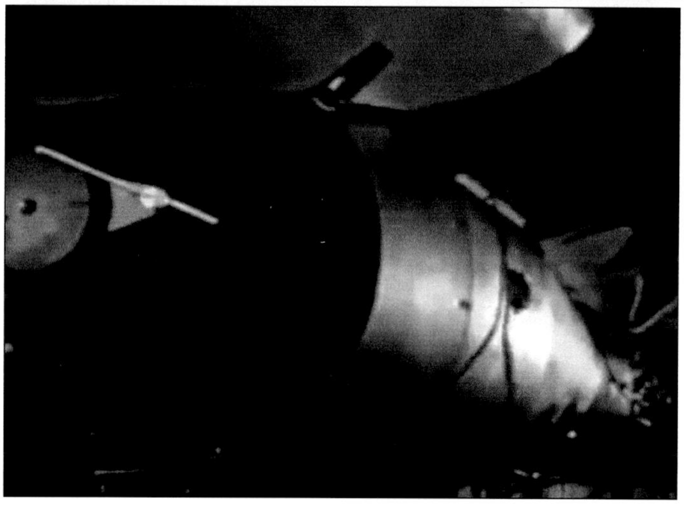

Brought by train to Oleny Air Base on the Kola Peninsula, the Tsar Bomba was prepared for the test and then hoisted into the heavily modified bomb bay of the Tu-95V.

Because of its sheer size (over eight metres in length) and weight (about 27,000kg), the casing of the RDS-202 – also designated the RN202 or AN602 – was colloquially referred to as the 'Tsar Bomba': this nickname was also used for some of its low-yield derivatives.

A view of the centre section of the specially modified Tu-95V destined to drop the Tsar Bomba. The aircraft received the white anti-flash finish on all undersurfaces and high up the sides, and up the entire rudder (for further details, see colour section).

The crew of pilot Lieutenant-Colonel A. E. Durnovtsev reporting its readiness to drop the biggest of all the bombs.

The bomb bay was enlarged to a length of 7.15 metres, with bulkheads at each end separating it from the rest of the fuselage's interior. Its width was 1.75 or 1.78 metres, depending on the source and so the Tsar Bomba could be carried. However, there was a penalty for this since the internal volume taken up by the enlarged bomb bay meant that there was less space for fuselage tanks which effectively reduced the amount of fuel and thus the aircraft's range.

It was decided to install a beam-type bomb rack similar to the BD-206 used on the Tu95K for the Kh-20 missile. The device fitted to the Tu-95V was designated BD7-95-242 or BD-242 for short; it incorporated three Der 5-6 bomb shackles, each capable of supporting a maximum load of 8,167kg. Yet the really difficult part was not to build the bomb rack itself but to ensure simultaneous load separation from all of its attachment points. In order to facilitate this, the release mechanism underwent bench tests on the ground until the desired synchronisation was achieved. Once the problem of releasing the payload was solved the whole apparatus was fitted inside the Tu 95V's bomb bay, attached to the load-bearing beams of the aircraft's structure. Apart from the bomb, the aircraft also had a defensive armament of six 23mm cannons in three twin mountings: a dorsal turret DT-V12, a ventral turret DT-N12 and a tail mounting DK-12, with the overall ammunition supply being 2,500 rounds.

Other than the enlarged bomb bay with its special fittings, and the reduced fuel capacity, there were hardly any the differences between the Tu-95V and a standard production aircraft – after all the Tu-95V was modified from one. However, the difference in the amount of fuel carried was not a minor issue as it reduced range and thus negated the ability of the Tu-95V to operate as a true strategic bomber.

The next stage of the program was to conduct tests which were carried out in September of the year 1959 at the Zhukovsky airbase under the supervision of Colonel S.M. Kulikov. These tests were successfully completed but the bomb was not to be dropped for the time being. Politics prevented a live test of the Tsar Bomba, for the Soviet Union's leader Khrushchev visited the United States and it was deemed to be undiplomatic to detonate a powerful thermonuclear device while there was a thaw in relations between the world powers. As a result, the Tu-95V was for the time being assigned to Uzyn Air Base, located near Belaya Tserkov in Ukraine. Since its bomb bay and bomb rack were configured for one-of-a-kind ordnance it was incapable of carrying a standard war load, and hence its usefulness was limited to aircrew flight training. But the Tu-95V was not to be employed in that role indefinitely.

The Biggest of Them All

As soon as the relationship between the Soviet Union and the United States took a turn for the worse the test of the Tsar Bomba was again on the agenda. However, before that unique piece of ordnance was dropped the specially modified Tu-95 flew a number of missions for the purpose of familiarisation and training. Notably on 20 September 1961 it accompanied a Tu-16 which performed a thermonuclear weapons test, though this was not a good day for the Tu-95 as it developed engine trouble and had to turn back while the Tu-16 flew on to complete its mission. The next sortie, on 6 October 1961, went much better and on that day the special Tu-95 piloted by Lieutenant-Colonel Durnovtsev successfully dropped a thermonuclear bomb of 4Mt yield on target coordinates 74.3°N and 51.6°E. Thus, the readiness of the aircraft and crew for the special mission was practically demonstrated.[3]

Finally, the test of the Tsar Bomba was scheduled to take place on a nuclear testing area at Cape Sukhoy Nos in the Novaya Zemlya archipelago on 30 October 1961. The operation was conducted from Oleny airfield on the Kola Peninsula where the Tsar Bomba was delivered by train. It had a weight of 24 tons to which 800kg more had

About to take off – the Tupolev warming its engines with its one-of-a-kind ordnance protruding from the bomb bay.

addition it also received an anti-flash white finish. The aircraft was captained by Lieutenant-Colonel Vladimir Fodorovich Martynenko.

As planned the Tu-95V, accompanied by the Tu-16, took off from Oleny airfield in the morning hours of 30 October 1961, setting course for Sukhoy Nos. At 11.33 Moscow time the aircraft arrived over the test area at an altitude of 10,500m (the Tu-16 at a somewhat different one to facilitate better filming of the bomb drop), and the Tu-95V released its load. After 188 seconds – the descent being slowed down by the breaking parachute – the bomb went off at an altitude of 4,200 metres (4,000 metres above ground) the coordinates of the detonation point being most frequently given as 73⁰51'N and 54⁰30'E. Lead tampers dialled down the yield of the bomb's calculated 100 megatons to circa 50 megatons (some sources put the actual yield at 58 megatons), but that was still enough to bring about a gargantuan explosion with a humongous fireball and mushroom cloud – 64km in height and with a diameter of almost 40km – as well as a very destructive shock wave. The latter destroyed all buildings in Severny village located 55 kilometres from the point of detonation, inflicted lesser damage for hundreds of kilometres away and was literally felt worldwide: the shockwave was registered by sensors to have travelled around the globe three times over!

The Tu-95V was 39km away from the point of detonation when it occurred, and 115km when the shock wave caught up with it, the figures for the Tu-16 being 53.5km and 205km respectively. Both aircraft were roughly shaken by the blast with the Tu-95 dropping some 800 metres before the pilot regained control and their white anti-flash finish was scorched too. Yet despite this, and a temporary communications failure caused by the electromagnetic shock of the

to be added accounting for a retarding parachute system. This was considerably more than the earlier optimistic calculations showed but the Tu-95V was still able to take off with this load. Another problem arose due to the bomb's size as it measured 8 metres in length and 2.1 metres in diameter, which was too much to fit into even the enlarged bomb bay of the aircraft meant to carry it. This problem was solved by removing the bomb bay's doors so that the bomb would be carried semi-recessed into the fuselage rather than fully internally. As a final touch, the aircraft was given a white anti-flash paint finish. For the mission, the Tu-95V was crewed by the pilot Lieutenant-Colonel Andrey Yegorovich Durnovtsev, navigator Major Ivan Nikiforovitch Kleshch and Flight Engineer Valenty Yakovlevitch Bruy.

During the nuclear test the Tu-95V was to be accompanied by a chase plane which was a standard production Tu-16A, number 3709. For this purpose, the Tu-16 was fitted with film-recording equipment as well as measuring instruments and hence designated as a flying laboratory (literally aircraft laboratory – *samolet laboratoriya*), in

Lieutenant-Colonel Andrey Yegorovich Durnovtsev, pilot of the Tu-95V that deployed the RDS-202, or the (first) Tsar Bomba.

Major Ivan Nikiforovich Kleshch, navigator of Durnovtsev's bomber.

Captain Valenty Yakovlevich Bruy, flight engineer of Dubrovtsev's Tu-95V.

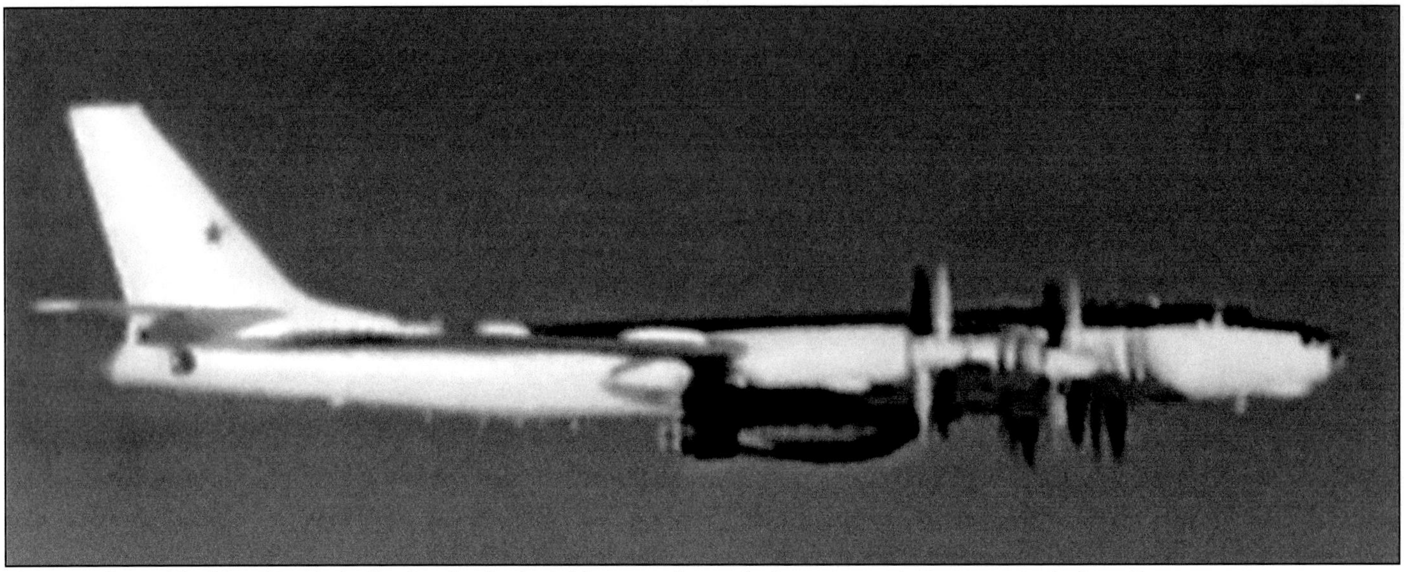

The specially modified Tu-95V, with the Tsar Bomba clearly visible, protruding under its centre section, en route to the 'target'.

A view less than a second after the drop of the Tsar Bomba, as the weapon began streaming its braking parachute.

Millisecond later, the parachute has reached the end of its cords, and began to open.

nuclear explosion, they flew back without further incident and landed safely.

Looking from Across the Iron Curtain

Khrushchev's boasting about a Soviet monster bomb did not escape US attention. Naturally the Americans desired to learn as much as possible about the new weapon. Since its test took place at Novaya Zemlya and not deep inside the Soviet Union's interior there was a realistic possibility to observe it from a relatively close position. For this purpose, an appropriately modified aircraft had to be readied at short notice. This was managed under the USAF's Project Big Safari, which provided management, direction and control of the acquisition, modification and logistics support for special purpose weapons systems derived from existing ones. A major advantage of Big Safari was that it facilitated operating outside normal acquisition procedures and therefore offered the ability to handle complex tasks at short notice. In this particular case a Boeing JKC-135A Stratotanker (serial 55-3127) which had previously been used for special systems testing was fitted with extra sensors and other modifications in just five days between 22-27 October 1961. Dubbed

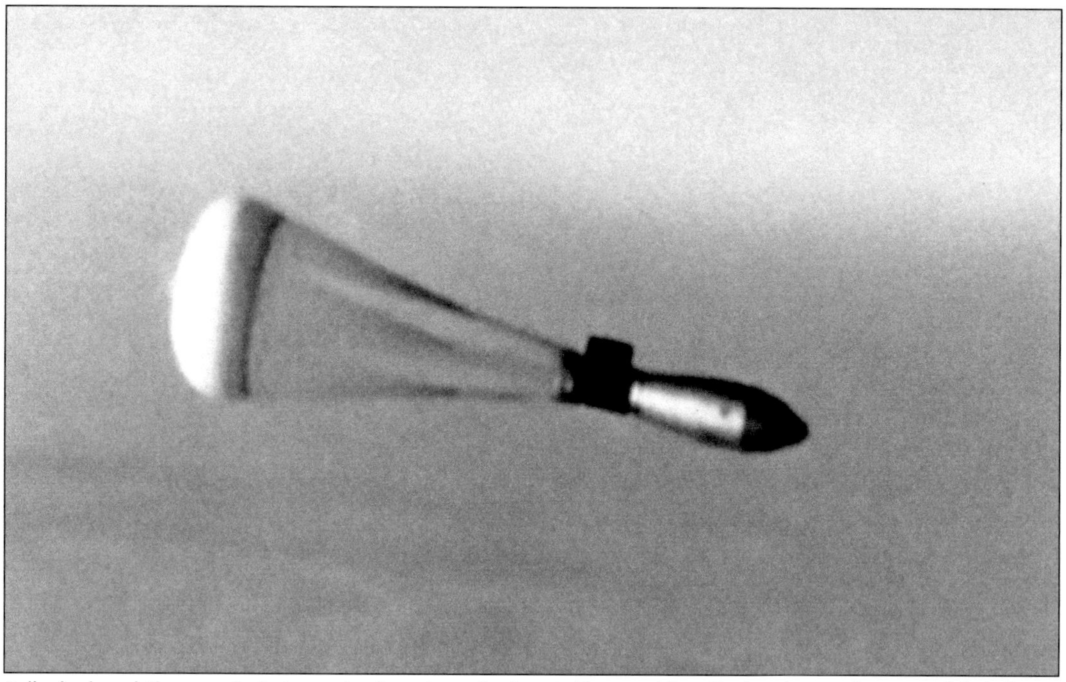

Fully deployed, the parachute began to slow-down the bomb, thus enabling the bomber to distance safely before the detonation.

carried and dropped by the same aircraft flown by the same pilot.

Despite already having gathered experience by dropping the Tsar Bomba, the Soviets still found it necessary to conduct two practice drops of inert bombs before making attempts with the live ones. Accordingly, the Tu-95, with A. Durnovtsev at the controls, accompanied by a Tu-16 flying laboratory piloted by V. Martynenko, as well as Tu-16 'double' backup aircraft flown by K. Lyasnikov, made a practice drop on 27 July 1962 and a second one on 2 August 1962. Being satisfied with the results of the practice drops, the Soviets went ahead with the dropping of a live high-yield thermonuclear bomb on 5 August 1962. On that day, the three aircraft – the Tu-95 with the bomb, a Tu-16 'flying laboratory' and a Tu-16 'double' – took to the air and set course for the Novaya Zemlya archipelago. The designated target could not be visually made out due to clouds, the upper layer of which was up to 8,000m, but this did not cause a problem because aiming was facilitated by the delivery aircraft's on-board radar. The weapon was released at an altitude of 10,500m and detonated at 3,500-3,600m, with the explosion's coordinates being given as 74.2°N and 52.5°E. The yield was circa 21Mt, this being the third most powerful thermonuclear explosion recorded, and the resulting mushroom cloud reached an altitude of 20km and a diameter of 35km. As far as it is known neither the aircraft involved nor their crews were seriously affected by the explosion. Yet another, and this time final, drop of a high-yield thermonuclear bomb was made on 24 December 1962 with the mission was again performed by the specially modified Tu-95 piloted by A. Durnovtsev. However, the necessary preparations were not without some problems as the bomb's breaking

Speed Light Bravo, the aircraft was able to observe the Tsar Bomba test (codenamed Joe-111 in the West), gathering important data in the process. This data was of considerable value to American nuclear scientists as well as the military and political leadership, with the US President John Kennedy showing his personal appreciation for the effort. As for *Speed Light Bravo*, it did not escape some scorching of its surfaces. More alarmingly it was also supposedly highly irradiated 'lighting up like a Christmas tree' as was commented at that time, though it was also claimed that no long term adverse health effects were experienced by its crew.[4]

More Megatons

Apparently, live testing the most powerful thermonuclear weapon ever was not satisfying enough for the Soviet leadership, as it was decided to conduct two additional such tests, though the bombs which were to be exploded had their yield still further reduced by more than half vis-à-vis the one detonated on 30 October 1961. It is also noteworthy that these thermonuclear devices were enclosed in a bomb casing of the same size and shape as the Tsar Bomba. They would also be

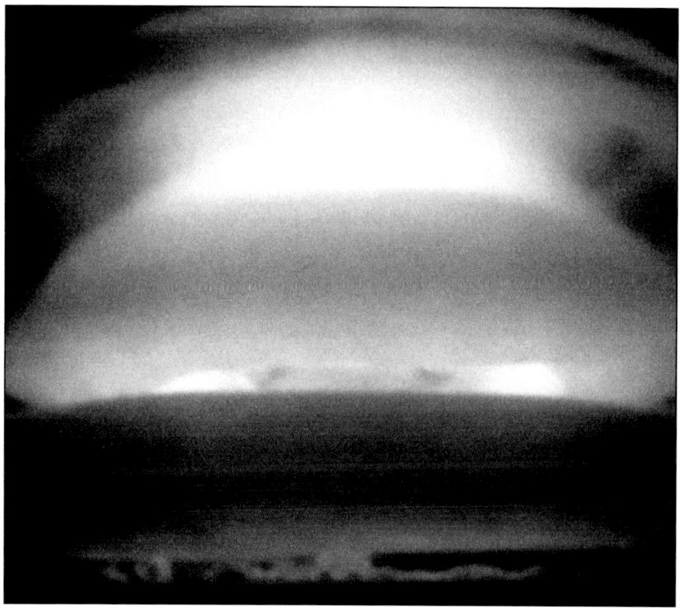

The Tsar Bomba test culminated with a detonation that created a giant fireball, the light of which was seen hundreds of kilometres away, and the power of which was not only felt thousands of kilometres away but threw the Earth some 0.2 degrees off its axis. As far as is known, this was the biggest man-made explosion ever.

Another casing of the RDS-202-type. Lieutenant-Colonel Durnovtsev was involved not only in the release of the Tsar Bomba but also piloted the same Tu-95 into two other test drops related to this type of weapon – though equipped with warheads of much lower-yield.

parachute had to be replaced which clearly demonstrated that such an undertaking could never be treated as just another routine task. Once all difficulties were dealt with the Tupolev was armed with the bomb and prepared for take off together with two 'flying laboratories'. In due time the go-ahead was given and the Tu-95 with the bomb, accompanied by a pair of Tu-16s tasked with recording the bomb's drop and explosion as well as measuring its aftermath's effects, took to the air. The trio set course for Novaya Zemlya where the target was located. As far as it is known the drop itself was accomplished without any noteworthy problems and the weapon detonated at an altitude of 3,750m at the coordinates 73.6°N and 57.5°E. The shockwave caught up with the bomber as it was 115km from the point of detonation, but it, as well as the accompanying 'flying laboratories', safely flew back to base. With a yield of 24.2Mt the test described was the second most powerful thermonuclear explosion to have ever taken place.[5] The event had a number of side effects which were outside thermonuclear weapons' development. For one thing a particularly intensive aurora borealis was observed which meant that the explosion must have affected the ionosphere. On a more practical side Norway and Finland issued an official note of protest to the USSR because the explosion's shockwave blew out many windows and doors in buildings located along the shores of the Barents Sea. It is also worthy to note that Andrey Durnovtsev was a man who not only flew the aircraft which dropped the three most powerful thermonuclear bombs but also took part in live testing nuclear weapons of lesser yield and was thus arguably the bomber pilot who dropped the most megatons ever.

Spectacular but Not Practical

Once atmospheric nuclear testing ceased, the Tu-95V was employed in the transport role ferrying elements of the Tu-144 supersonic airliner.[6] Subsequently the Tu-95V was assigned to the 1023rd Heavy Bomber Aviation Regiment based at Semipalatinsk where it was used for training before being relegated to ground instruction airframe status and finally scrapped in the mid-1980s.

While the aircraft which dropped the monster bomb rightfully deserved attention, the men who flew the Tu-95V on its historic Tsar Bomba mission should not be forgotten either. The pilot A. E. Durnovtsev and navigator I. N. Kleshch were bestowed with the Hero of the Soviet Union title receiving the Gold Star medal numbers 11127 and 11132 respectively. In addition, V. F. Martynenko who flew

A Myasishchev 3M bomber like this one, drawn from the 1096 TBAP, took part in a next-to-unknown test drop of a nuclear weapon over the Novaya Zemlya test site on 27 February 1962.

A Tu-16 nuclear bomber, its anti-flash finish on the fuselage sides and lower surfaces clearly distinguishable, undergoing open-air maintenance in wintery scenery. While as far as it is known the image itself was not taken during the nuclear and thermonuclear bombing spree of 1961-1962, the servicing of Tupolev bombers before or after such missions in the far north would have looked very much like this.

Painted in similar fashion was this Tu-95 bomber, intercepted by NATO fighters over the Norwegian Sea. Notable is not only its anti-flash livery down the entire fuselage and undersurfaces of the wing, but also the large 'bort' 75 applied on the nose, and near the top of the fin. (Albert Grandolini Collection)

the Tu-16 chase plane was also awarded the same title and medal number 11131.[7] Martynenko, it may be noted, was also involved in the preparations for the first Soviet thermonuclear bomb drop test.

Closing this matter, it is worth noting that while the Tsar Bomba and its reduced yield derivates were certainly very powerful, their practical value was limited. Their size meant that they could only partially fit into the enlarged bomb bay of a modified Tu-95, at the expense of internal fuel tanks which in turn cut the aircraft's range, thus negating its ability to operate in a truly strategic role.

An underside view of a Tu-16K-10 bomber: notable under the centreline is a K-10S turbojet powered cruise missile. Four sub-variants of cruise missile-carrying Tu-16s emerged, including the Tu-16KS (capable of carrying two KS-1s), Tu-16K-10 (which could be armed with one K-10S, or one K-10S and two KSR-2s or KSR-5s), and Tu-16K-11-16 (armed either with two KSR-2s or two KSR-11s).

A trio of Tu-95Ks, all armed with huge Kh-20 cruise missiles installed under the centre fuselage, photographed during a parade in Moscow in the 1960s. (Albert Grandolini Collection)

A closer view of a Tu-95K with a Kh-20 in its position. (Albert Grandolini Collection)

5
STORM BEFORE THE CALM

Tests of the Tsar Bomba and its reduced yield derivates were unique as well as spectacular and thus deserved a separate description. That said, during 1961-62, a number of other crucially important developments took place concerning Soviet nuclear and thermonuclear weapons and their primary arm of delivery; Soviet Long-Range Aviation (*Dalnaya Aviatsiya*, DA). Indeed, during this period and with few interruptions, the Soviets were – just like the Americans – running live testing of nuclear weapons on at least a weekly – and frequently on a daily – basis. It was also one of the high points of the Cold War culminating with the Cuban Missile Crises in October 1962. Thus, not surprisingly, the years mentioned were intense as far as nuclear and thermonuclear weapon testing was concerned, with dozens of live-weapon air-drops. These missions were usually launched from Oleny airfield on the Kola Peninsula with the targets located in the Novaya Zemlya archipelago. For the most part these were performed by Tu-16 bombers though occasionally other types were utilised as well. For example, on 27 February that year a Myasishchev M3 form the 1096 TBAP, with A. N. Belenkov at the controls and I. I. Nasedkin as navigator, dropped a 320kt yield bomb. This was a somewhat little-known nuclear test and rarely mentioned.[1] Among the many tests conducted by the Soviets in 1961-62 several in particular are worthy of note. Leaving aside those already described above these certainly include trials of air-launched missiles with nuclear warheads as well as Soviet tactical preparations for thermonuclear Armageddon.

An overhead view of a Tu-16K (construction number 1883820) of the Soviet Naval Aviation. Notable is the prominent nose-bulge, containing a radar necessary to detect targets for its cruise missiles. (US DoD)

A view under the left wing of a Tu-16 of the Soviet Naval Aviation, milliseconds after the release of a KSR-2 cruise missile (ASCC/NATO-code 'AS-5 Kelt'). (via Tom Cooper)

The Missileries

While thus far only the delivery of air dropped bombs from manned aircraft has been discussed, other tools for nuclear attack were also being developed. Towards the end of the Second World War Luftwaffe Heinkel He 111 bombers were launching V-1 'flying bombs' at southern England, thus demonstrating the practical feasibility of air-launched land-attack cruise missiles. The Soviets recognised the potential offered by such weapons early on and pursued their development from the late 1940s. The general layout adopted was that of a small sized aircraft and accordingly the first Soviet cruise missiles

To Quench the Thirst

The Russians thought of extending aircraft range by means of in-flight refuelling as early as 1917 but at that time the idea failed to materialise. During the interwar years the Soviets reached the practical stage of actually performing in-flight refuelling, yet these were mere experiments which had little value beyond demonstrating that transferring fuel from one aircraft to another while both were in flight was possible. When war came, research and trials in this field, as well as a number of others, were suspended. Resources had to be devoted to work which was deemed of greater immediate practical value for the war effort. It was only after the Second World War that the Soviets could start serious work on an in-flight refuelling capability.

Early post-war experiments involved, among others, a pair of US-made North American B-25 Mitchel bombers, one being the aerial tanker and the other refuelling from it, as well as a Yak-15 jet fighter receiving fuel from a Tu-2 bomber. However, it was only due to the need to fulfil the requirements of the Long-Range Aviation that the real breakthrough for Soviet in-flight refuelling came.

Aircraft range is limited by factors such as a given type's fuel consumption vs. its fuel capacity and Soviet bombers were no exception. In addition, since they were based in the Soviet Union, reaching targets in Western Europe would be difficult, while those in North America impossible. Thus, it was imperative to provide the Soviet bomber force with means to extend its range and in-flight refuelling offered just that. The first post-war modern Soviet long-range heavy bomber, the Tu-4 was also the first Soviet aircraft in this category to be converted for air-to-air refuelling. However only six Tu-4s were appropriately modified: three were converted to aerial tankers and the remaining trio was fitted for receiving fuel in-flight. While this was obviously not enough to give the Soviet bomber force the needed extended range it was nevertheless an important step in this direction. In fact, experience gained with Tu-4 aerial refuelling directly led to the development of the Tu-16 aerial tanker.

It was the Tu-16 *Zapravshchik* (tanker), which gave the Soviet Long-Range Aviation and Naval Aviation a real air-to-air refuelling capability and thus a considerably extended range.[15] Dozens of the Tupolevs were converted for this role, though exact numbers differ from source to source ranging from just 46 up to 114. The *Zapravshchik* could carry up to 24,500kg of additional fuel which would be transferred wingtip-to-wingtip via a flexible hose. In-flight refuelling operations were conducted in such a manner that the tanker and the receiving aircraft would assume an echelon formation with the former somewhat ahead. The tanker deployed the fuel transfer hose from its right wing while the receiving aircraft manoeuvred so as to have the hose slide into an appropriate fitting under its left wing. The principal advantage of the method described lay in that the aircraft which was being refuelled was not in the aerial tanker's wake, thus avoiding the jet stream and vortexes of the latter. In addition to the Tupolevs, a number of Mayashchevs – which had proved unsatisfactory in the bomber role – were also converted to aerial tankers utilising a probe and drogue system for fuel transfer. Soviet bombers' jet engines were fuel thirsty and the successful development of the in-flight refuelling capability described did much to quench it thus considerably alleviating – though not fully solving – the issue of their limited range.

One of the Myasischev 3MDs (ASCC/NATO-codename 'Bison C'): originally designed as a cruise missile carrier, only nine examples of this variant were ever manufactured. Eventually, most were converted to 3MS tanker aircraft instead. (US DoD)

were designated as *Samolety-snaryady* – literally *aircraft missiles* – but in fact meant aircraft *as* missiles, for indeed these were in essence unmanned kamikaze jets. Subsequently the designation *Krylata raketa* was adopted, meaning winged rocket, though the 'rocket' would be best substituted with 'missile' in the English language. The Soviets developed a whole range of such weapons including air-launched land-attack missiles with both conventional as well as nuclear warheads, missiles homing onto radar emissions, and anti-ship missiles, again with conventional and nuclear warheads. It should be noted that Soviet cruise missiles and anti-ship cruise missiles were closely related, as they shared much in airframe and propulsion design but differed in guidance methods.[2] The first Soviet weapon system of this kind to reach operational status was the KS-1 Kometa (AS-1 Kennel) initially carried by an appropriately modified Tu-4 bomber. Soon enough a number of other types of air-launched cruise and anti-ship missiles were fielded.

Air-launched cruise missiles were within a relatively short time mated with nuclear warheads to give an attacking aircraft the capability to conduct a nuclear strike from a stand-off range. With new bomber types such as the Tu-16 jet and Tu-95 turboprop

Frequently shown in public while unleashing unguided rockets, the primary purpose of the Sukhoi Su-7B fighter-bomber – four of which are seen in this image – was nuclear strike. The primary targets for Soviet Su-7-equipped units deployed in East German and, later on, Czechoslovakia, were NATO's early warning radar sites and MIM-23 HAWK SAM-sites in Western Germany. (via Tom Cooper)

becoming operational, these aircraft were also adopted as missile launch platforms (*raketonosets*) with underwing pylons or attachment points for semi-recessed carriage in the bomb bay. According to the majority of sources the Soviets live tested an air-launched cruise missile with a nuclear warhead for the first time on 8 October 1961, the test being codenamed Joe-99 in the West. In the course of this test a K-10S (AS-2 Kipper) missile was launched from a Tu-16 at a distance of 100km form the target located on Cape Chernyy in the Novaya Zemlya archipelago.[3] Other sources give the target's location as either Chyornaya Guba Bay[4] or Cape Sukhoy Nos[5] being in agreement that it was an airburst of 15kt yield. Some Russian sources seem to indicate an earlier test of this kind, however. *Atom* magazine states that a KS-7 missile with a nuclear warhead was tested at Semipalatinsk on 17 January 1958, being launched by an unspecified aircraft from an altitude of 8,000m and detonating at 500m.[6] It would be an interesting revelation showing that the Soviets had air-launched cruise missiles with nuclear warheads earlier than it is usually assumed. However, the KS-7 missile was not an air-launched weapon but another and less commonly used designation of the FKR-1 (SSC-2B Samlet) ground-launched cruise missile and seriously compromises the credibility of the presented claim. Sources usually used for reference of nuclear testing state that the 17 January 1958 test was that of a low-yield (0.5kt) nuclear bomb.[7]

In addition to the aforementioned trials, Mozdok based Tu-95s of the 182 TBAP would test launch 19 Kh-20 (AS-3 Kangaroo) cruise missiles in the time period between January and October 1962. Out of these, 15 struck their intended targets proving that the weapon system though not failproof showed a degree of reliability which should have been sufficient for operational use. While none of the Kh-20s fired carried a nuclear or thermonuclear warhead, this missile type could be fitted with one of 3Mt yield. Thus, the Soviet bomber force's capability for thermonuclear attack by means of air-launched cruise missiles was clearly demonstrated.

The Carrier Killer

While little can be said with certainty regarding the first Soviet test of an air-launched nuclear-tipped cruise missile, much more is known about their first test of an anti-ship cruise missile with a nuclear warhead. To give some context to the latter it needs to be pointed out that one area in which the West, especially the United States, decidedly had superiority over the Soviet Union was naval power. In particular, US Navy aircraft carriers had capabilities unmatched by the Soviets. The Soviets were in dire need of finding ways of neutralising the US carriers, preferably with a single powerful blow. Guided missiles with a nuclear warhead offered just that, and for this reason it should be no surprise that the Soviets found this combination to be an attractive option. Accordingly, such weapons were developed and the Soviet Navy wanted them live tested. Thus, the test of a nuclear-tipped anti-ship missile was planned for 1962 but nature intervened as floating ice destroyed the practice target before the first attempt could be made. This proved only a temporary setback and conditions would permit the live firing of an anti-ship missile with a nuclear warhead scheduled for 22 August 1962. Rather unsurprisingly, considerable importance was attached to this, with Admiral V. A. Kasatonov – Commander of the Soviet Northern Fleet – being in charge of the undertaking codenamed 'Shkval' ('squall'). A target fitted with radar reflectors was floated in Bashmachnaya Guba Bay in the Novaya Zemlya archipelago.[8] It was engaged with a K-10S (AS-2 Kipper) or a KSR-2 (AS-5 Kelt) anti-ship cruise missile launched from a Tu-16K bomber. Reportedly the Tupolev's crew, commanded by V.F. Krypiakov (at that time he either held the rank of a Major[9] or was a Lieutenant Col and the CO of the 924th Guards MRAP[10]), found the task very unnerving and fumbled the engagement procedure on the first attempt. The target was acquired by the aircraft's on-board radar from a range of 400km and subsequently the missile was launched against it. The missile guided well, with the 6kt warhead detonating 60m above the water surface at the coordinates 70°42'0" N and 54°36'0" E, resulting in the target's obliteration. Soviet Naval Aviation could now lay claim to possessing a tested weapon of great destructive power in its arsenal.[11]

Practicing for Doomsday

As much as air-launched missiles with nuclear warheads were a welcome addition to the Soviet arsenal, the aerial bomb still remained the primary means of nuclear attack, in particular if a weapon of large yield was to be delivered on target. It also needs to be pointed out that the test drops of nuclear and thermonuclear weapons were not an end to themselves, neither was the manufacture of the bombs and the aircraft which carried them. Instead, it all served the purpose of forging a workable arsenal which if need be could be put to actual use. Fortunately for all concerned this never came, but the tools of long-range nuclear air attack still had to be tested. This, however, was not in the fashion of a development trial but in a way more resembling an actual wartime mission.

Relatively large-scale manoeuvres of the Long-Range Aviation involving actual drops of live thermonuclear bombs were planned.[12] Officially these were designated as *Letno-Takticheskiye Ucheniya* (LTU) (Flying Tactical Exercises)[13]. However, the attack for which the delivery tactics were practiced would not be tactical at all but in fact be of long-range strategic nature. To summarise the whole affair in a succinct manner: the Soviets were to practice for doomsday. The undertaking was managed by a commission headed by General I. L. Turkel who at that time held the post of the deputy commander of the Long-Range Aviation.

The exercises were planned for September 1962 and were to be staged form the Oleny airfield on the Kola Peninsula. Accordingly, a number of Tu-16 bombers from operational units – that is the 132 TBAP and the 185 TBAP – were brought together there, as were several Tupolevs fitted with measuring instruments (the 'flying laboratories'). The exercises took place on 15 and 16 September 1962. On each day, 24 Tu-16s were involved, including three 'flying laboratories', and each time a single live thermonuclear bomb was dropped from an altitude of 11,300m against targets located on a nuclear testing area at Cape Sukhoy Nos in the Novaya Zemlya archipelago. In both cases the bombs were set to detonate at an altitude of 2,200–2,225 metres with the point of explosion being given as 74.4^{0}N and 51.5^{0}E for the first, and 74.2^{0}N and 51.6^{0}E for the second, while the yields were estimated to be 3.1Mt and 3.25-3.3Mt, respectively. The bomb drops were each time performed by aircraft and crews of the 185 TBAP: Pilot Lieutenant-Colonel A. A. Popov with Navigator Major Vinogradov, on the first day and Lieutenant Col I. Bazhenov (the unit's deputy commander) on the second. The aircraft involved sustained some scorching of surfaces and damage to external fittings, requiring repairs, but these were not of a really serious nature. Similarly, the bombers' crews were not seriously affected either. Once the exercises were complete, it was concluded that the Long-Range Aviation was ready for the combat use of nuclear weapons – a chilling statement indeed, for it meant that the Soviets saw themselves as ready for nuclear war which, if it ever came, would be nothing other than a catastrophe for humanity.

Fortunately, the Soviet bomber crews' proficiency in nuclear weapon delivery was not to be tested in actual combat, but it came very close to that and this in a relatively short time after the manoeuvres described above. During the Cuban Missile Crisis of October 1962, units of the Long-Range Aviation were put on alert, with preparations being made to launch bombers with a thermonuclear war load should

Table 1: Composition of the 2nd Independent Heavy Bomber Aviation Corps, 1961

Unit	Aircraft Type	Base
13-ya Tyazholaya Bombardirovochnaya Aviatsionnaya Diviziya **(13th Heavy Bomber Aviation Division)**		Poltava
184-y Tyazhelyy Bombardirovochnyy Aviatsionnyy Polk (184th Heavy Bomber Aviation Regiment)	Tu-16	Priluki
185-y Tyazhelyy Bombardirovochnyy Aviatsionnyy Polk (185th Heavy Bomber Aviation Regiment)	Tu-16	Poltava
225-y Tyazhelyy Bombardirovochnyy Aviatsionnyy Polk (225th Heavy Bomber Aviation Regiment)	Tu-16	Poltava
15-ya Tyazholaya Bombardirovochnaya Aviatsionnaya Diviziya **(15th Heavy Bomber Aviation Division)**		Zhitomir
250-y Tyazhelyy Bombardirovochnyy Aviatsionnyy Polk (250th Heavy Bomber Aviation Regiment)	Tu-16	Strvi
251-y Tyazhelyy Bombardirovochnyy Aviatsionnyy Polk (251st Heavy Bomber Aviation Regiment)	Tu-16	Belaya Tserkov
341-y Tyazhelyy Bombardirovochnyy Aviatsionnyy Polk (341st Heavy Bomber Aviation Regiment)	Tu-22	Ozernoe
106-ya Tyazholaya Bombardirovochnaya Aviatsionnaya Diviziya **(106th Heavy Bomber Aviation Division)**		Uzin
182-y Tyazhelyy Bombardirovochnyy Aviatsionnyy Polk (182nd Heavy Bomber Aviation Regiment)	Tu-95	Mozdok
409-y Tyazhelyy Bombardirovochnyy Aviatsionnyy Polk (409th Heavy Bomber Aviation Regiment)	Tu-95	Uzin
1006-y Tyazhelyy Bombardirovochnyy Aviatsionnyy Polk (1006th Heavy Bomber Aviation Regiment)	Tu-95	Uzin
199-y Otdelnyy Dalnerazvedyvatelnyy Aviatsionnyy Polk **(199th Independent Long-Range Reconnaissance Aviation Regiment)**		Nezhin

it come to war. Knowing the disposition of forces, and in particular the number of strategic nuclear weapons delivery systems at that time, the Soviets could not have hoped for victory. Having said this, over the coming years the Soviet nuclear arsenal grew both numerically and capability-wise in particular, with the build-up of an intercontinental ballistic missile force.

On the Lighter Side

While the Soviets were fielding ever-more heavy bombers and thermonuclear bombs to arm them, they did not neglect the other end of airpower's spectrum. In the early 1960s a number of then-modern tactical fighters and fighter-bombers were also introduced into service, in particular the MiG-21 and Su-7. It was obvious that tactical airpower was also in need of a nuclear strike capability which in turn facilitated the need for an appropriate weapon to be developed for that purpose. This materialised in the form of the 244N nuclear bomb which could be carried on external hard points of tactical combat aircraft flying at supersonic speeds.

Work on the bomb culminated in a successful live test carried out on 27 August 1962 at Semipalatinsk.[14] The bomb exploded at the coordinates 50.4°N and 77.8°E with the detonation having a yield of 11kt. The actual drop was performed by Lt. Col. A. I. Shein who set at the controls of a Su-7B fighter-bomber. To be precise, the bomb was not so much dropped as in fact tossed; the bomb delivery method was such that the aircraft would release the war load while climbing at about 44–50 degrees. Upon separating from the aircraft's hardpoint the ordnance would continue on an arching trajectory.

Having been successfully tested, the 244N nuclear bomb was put into operational service. Manufactured in several variants differing in yield as well as other characteristics, for many years it constituted an important part of Soviet tactical nuclear arsenal. In addition it is also noteworthy that the IAB-500 imitation bomb had a casing of similar size and shape and was meant to imitate the explosion of a nuclear bomb during military manoeuvres. Filed with a mixture of liquid petroleum products and phosphorous, the bomb would be set off by a small high explosive charge. The detonation produced a large fireball which would subsequently turn into a mushroom cloud. While obviously several orders of magnitude smaller than an actual nuclear explosion the visual effect was still spectacular.

Final Fireworks

The last tests of the sort described thus far were performed by the Soviets on 25 December 1962. That day the Soviets dropped two bombs: a thermonuclear one with a yield of 3.1Mt and a nuclear one of 8.5kt. Their points of detonation were 73.4°N and 56.5°E and 73°N and 55°E respectively, both locations being in the Novaya Zemlya archipelago. These were the final Soviet live tests of air dropped thermonuclear and nuclear bombs as well as their last atmospheric tests. The next year, on 5 August 1963, the *Treaty Banning Nuclear Weapon Tests in the Atmosphere, in Outer Space and Under Water* was signed in Moscow by the United States, the Soviet Union and the United Kingdom. With the signing, all tests of nuclear weapons were banned except those conducted underground. Thus, the nuclear and thermonuclear bombing spree of 1961-1962 was a storm before the ensuing calm.

6
MOSCOW'S MACE

By 1962 atmospheric nuclear testing came to an end, which meant that no more live nuclear bombs would be dropped by Soviet bombers. Arguably, this was no longer necessary for by that time the main purpose of the testing – which was to provide the Soviet Long-

A Tu-16 bomber under a camouflage net, with another about to land in the background. Their anti-flash white undersides reveal both of them as Tu-16A nuclear-armed bombers. The Soviets were no strangers to the idea of dispersing and camouflaging their bomber-fleet and this was frequently practiced during exercises.

Table 2: Composition of the 6th Independent Heavy Bomber Aviation Red Banner Corps, 1961

Unit	Aircraft type	Base
22-ya Tyazholaya Bombardirovochnaya Aviatsionnaya Diviziya **(22nd Heavy Bomber Aviation Division)**	Bobruysk	
121-y Tyazhelyy Bombardirovochnyy Aviatsionnyy Polk (121st Heavy Bomber Aviation Regiment)	Tu-16	Machulishche
200-y Tyazhelyy Bombardirovochnyy Aviatsionnyy Polk (200th Heavy Bomber Aviation Regiment)	Tu-16	Bobruysk
203-y Tyazhelyy Bombardirovochnyy Aviatsionnyy Polk (203rd Heavy Bomber Aviation Regiment)	Tu-22	Baranovichi
55-ya Tyazholaya Bombardirovochnaya Aviatsionnaya Diviziya **(55th Heavy Bomber Aviation Division)**	Migalovo	
45-y Tyazhelyy Bombardirovochnyy Aviatsionnyy Polk (45th Heavy Bomber Aviation Regiment)	Tu-16	Migalovo
173-y Tyazhelyy Bombardirovochnyy Aviatsionnyy Polk (173rd Heavy Bomber Aviation Regiment)	Tu-16	Migalovo
52-y Tyazhelyy Bombardirovochnyy Aviatsionnyy Polk (52nd Heavy Bomber Aviation Regiment)	Tu-16	Shaykovka
201-ya Tyazholaya Bombardirovochnaya Aviatsionnaya Diviziya **(201st Heavy Bomber Aviation Division)**	Engels	
1096-y Tyazhelyy Bombardirovochnyy Aviatsionnyy Polk (1096nd Heavy Bomber Aviation Regiment)	3M	Engels
1230-y Tyazhelyy Bombardirovochnyy Aviatsionnyy Polk (1230th Heavy Bomber Aviation Regiment)	3M	Engels
326-ya Tyazholaya Bombardirovochnaya Aviatsionnaya Diviziya **(326th Heavy Bomber Aviation Division)**	Tartu	
132-y Tyazhelyy Bombardirovochnyy Aviatsionnyy Polk (132nd Heavy Bomber Aviation Regiment)	Tu-16	Tartu
402-y Tyazhelyy Bombardirovochnyy Aviatsionnyy Polk (402nd Heavy Bomber Aviation Regiment)	Tu-16	Balbasovo
840-y Tyazhelyy Bombardirovochnyy Aviatsionnyy Polk (840th Heavy Bomber Aviation Regiment)	Tu-16	Soltsy
290-y Otdelnyy Dalnerazvedyvatelnyy Aviatsionnyy Polk (290th Independent Long-Range Reconnaissance Aviation Regiment)	Tu-22R	Zyabrovka
179-ya Otdelnaya Aviatsionnaya Eskadrilya (179th Independent Aviation Squadron)	M-4	Šiauliai

Range Aviation with workable weapons of great destructive power – was attained. These weapons would remain nothing but oversized firecrackers without effective means of delivery. While the aim of the volume is not to describe in detail the Soviet bomber types, as this has been comprehensibly done in a number of easily accessible monographes, one cannot avoid devoting a few words to the Soviet long-range bomber force. The *Dalnyaya Aviatsiya* (DA) – Long-Range Aviation – reached its peak in 1961, by when it was organised into three aviation corps, these being:

- the *2-y Otdelnyy Tyazholyy Bombardirovochnyy Aviatsionnyy Korpus* (OTBAK) (2nd Separate Heavy Bomber Aviation Corps)
- the *6-y Otdelnyy Tyazholyy Bombardirovochnyy Aviatsionnyy Krasnoznamonnyy Korpus* (6th Separate Heavy Bomber Aviation Red Banner Corps)
- the *8-y Otdelnyy Tyazholyy Bombardirovochnyy Aviatsionnyy Korpus* (8th Separate Heavy Bomber Aviation Corps).

The three corps were headquartered in Vinnitsa, Smolensk and Blagoveshchensk respectively.[1] During the time period in question the DA was commanded by Marshal Vladimir Alexandrovich Sudets who oversaw thermonuclear weapons and guided missiles, as well as jet and turboprop bombers entering service of the force under his command. In 1962 Sudets was replaced by Marshal Filip Aleksandrovich Agalcov.[2]

By this time, the bomber type forming the backbone of the Long-Range Aviation was the Tu-16, 1,507 were built up until 1963 when production ended. Out of these aircraft, about a half went to the Long-Range Aviation with the other half going to the Naval Aviation. Another important type was the Tu-95, which remained in production considerably longer. By December 1962 some 78 had been delivered, including those fitted for missile carriage and launch.[3] In addition, 93 Mayashchev M-4 and M3 bombers, the Soviet's great unfulfilled hope, were built.[4] Finally the Long-Range Aviation also had the supersonic Tupolev Tu-22 in its inventory; of the 311 (excluding two prototypes) manufactured some 55 would have been delivered by 1962. The main

Table 3: Composition of the 8th Independent Heavy Bomber Aviation Corps, 1961[5]

Unit	Aircraft type	Base
55-ya Tyazholaya Bombardirovochnaya Aviatsionnaya Diviziya **(55th Heavy Bomber Aviation Division)**	Vozdvizhenka	
303-y Tyazhelyy Bombardirovochnyy Aviatsionnyy Polk (303rd Heavy Bomber Aviation Regiment)	Tu-16	Zavitinsk
444-y Tyazhelyy Bombardirovochnyy Aviatsionnyy Polk (444th Heavy Bomber Aviation Regiment)	Tu-16	Vozdvizhenka
55-ya Tyazholaya Bombardirovochnaya Aviatsionnaya Diviziya **(55th Heavy Bomber Aviation Division)**	Ukrainka	
40-y Tyazhelyy Bombardirovochnyy Aviatsionnyy Polk (40th Heavy Bomber Aviation Regiment)	M-4	Ukrainka
79-y Tyazhelyy Bombardirovochnyy Aviatsionnyy Polk (79th Heavy Bomber Aviation Regiment)	3M	Ukrainka
279-ya Tyazholaya Bombardirovochnaya Aviatsionnaya Diviziya **(79th Heavy Bomber Aviation Division)**	Semipalatinsk	
1023-y Tyazhelyy Bombardirovochnyy Aviatsionnyy Polk (1023rd Heavy Bomber Aviation Regiment)	Tu-95	Semipalatinsk
1226-y Tyazhelyy Bombardirovochnyy Aviatsionnyy Polk (1226th Heavy Bomber Aviation Regiment)	Tu-95	Semipalatinsk
1225-y Tyazhelyy Bombardirovochnyy Aviatsionnyy Polk (1225nd Heavy Bomber Aviation Regiment)	Tu-16	Belaya
1229-y Tyazhelyy Bombardirovochnyy Aviatsionnyy Polk (1229nd Heavy Bomber Aviation Regiment)	Tu-16	Belaya
219-y Otdelnyy Dalnerazvedyvatelnyy Aviatsionnyy Polk (219th Independent Long-Range Reconnaissance Aviation Regiment)	Tu-16	Spassk-Dalniy

A view at the left side of the spacious Tu-95's cockpit, with its pilot at the controls. (Albert Grandolini Collection)

The Need for Speed

Within a relatively short time from the advent of the jet age, the sound barrier was breached and soon even higher speeds were attained by aircraft. Thus, it seemed that fielding a supersonic bomber was technically within reach, and the Soviet designers went to work on such.

In particular, Myasishchev's efforts produced interesting results in the form of the M-50 (ASCC reporting name 'Bounder') design. With a long and slender fuselage and delta wings it seemed to be able to cut through the air with great speed over an operating range of around 10,000 kilometres. Furthermore, the original idea was to equip it with the long-range M-61 supersonic cruise missiles developed by the same design bureau, which would enable it to strike targets from as far as 1,000 kilometres away – well outside the range of enemy air defences.

However, the M-50 airframe's development outpaced engine development. A prototype was built and test flown for the first time on 27 October 1959, with two VD-7 engines installed on underwing pylons, and two VD-7B engines installed on wingtips: these proved unable to accelerate the new bomber at speeds higher than 1,050km/h or Mach 0.99 – and then only in a shallow dive. Since the aircraft was unable to meet the requirements it was built for, further work was discontinued in 1960 and the Myasischev Design Bureau closed. The M-50 was flown only 19 times, the last instance being during the Tushino Air Show on 9 July 1961. Escorted by two MiG-21 fighters, the sleek aircraft looked like an embodiment of cutting edge aviation technology, while it was in fact a failure. For all the attention the M-50 attracted, its public appearance was also its swan song.

Undeterred by difficulties with the M-50, the Myasischev Design Bureau had numerous other designs under development before it was closed. One was the M-52, which could be justifiably termed an 'improved M-50': though it suffered from similar problems to its predecessor. The aircraft underperformed while at the same time the Soviet Union's leader Nikita Khrushchev was pushing for ballistic missile development as the country's primary means of strategic nuclear weapons delivery. Further work was terminated, as was that on all other supersonic bombers. Ultimately, the first prototype of the M-52 was scrapped. The sole M-50 was only slightly more fortunate: it survives to this day as an open-air exhibit at the Central Air Force Museum at Monino Airfield. To give credit where it is due, one cannot help but to point out that Myasishchev's failure was not rooted in faulty airframe design, but rather resulted from the lack of engines with suitable performance.

As a side note, Tupolev's supersonic bomber, the Tu-22, actually reached operational service but even its creator was fully aware that the aircraft was far from perfect: indeed, it suffered from many shortcomings.

Conventional power plants seemed unable to give Soviet bomber aircraft the desired performance. However, harnessing nuclear power for the purpose of aircraft propulsion could potentially provide a solution to most problems. A nuclear-powered aircraft would have effectively unlimited range and great speeds could have probably been attainable as well. Yet putting the idea into practice was not an easy task. The Soviets decided to tackle it with a step-by-step approach, first starting by fitting a nuclear reactor on-board an aircraft. For that purpose, a standard production Tu-95 bomber was taken in hand and appropriately modified. A nuclear reactor was placed in the aircraft's bomb bay. It was for obvious reasons a compact design but still somewhat oversized for the aircraft, requiring aerodynamic fairings over the top and bottom of the fuselage to cover it. In addition, external tanks with reactor coolant were carried, one under each wing.

A pair of photographs showing the sole M-50 prototype, accompanied by two MiG-21s and thus nicely illustrating what a sleek – but also huge – design this was. Notable is the 'bort' 12: frequently repainted, this was used to confuse the Western intelligence regarding the number of M-50s 'in service'.

The nuclear reactor was not to power the Tupolev's engines or any of the aircraft's other systems, its sole purpose was to test the practical feasibility of installing a reactor on-board an aircraft. Apart from 'mating' the reactor with the aircraft, the next biggest challenge was to provide for the crew's safety. For this purpose, radiation shields were devised composed of lead, polyethylene and resigns. The aircraft thus modified was designated Tu-95LAL.

Between May and August 1961, the specially modified Tupolev made 34 flights with the reactor both 'cold' as well as 'hot', though the latter only on a few occasions. While the feasibility of a nuclear reactor-aircraft combination was technically proven, and the radiation shielding also seemed satisfactory, the program was discontinued. It was clear that building an aircraft that would not only carry a nuclear reactor but would actually be powered by such was a technically difficult task which required both a lot of time and resources. Since it became known that the Americans had abandoned their nuclear-powered aircraft projects there was no more incentive for the Soviets to continue with an analogous design of their own. Thus, the Tu-95LAL was grounded and subsequently broken up while other design bureaus ceased work on their nuclear-powered aircraft projects, such as the Myasishchev M-60 which never took off from the drawing board.

The late 1950s and early 1960s was a time period when Soviet aircraft designers tried a number of concepts to build strategic bombers with unrivalled performance in speed and endurance. To achieve this, many ideas, including unorthodox ones, were tested, some of which proved to be dead ends. Arguably the principal reason for those failures was the limits of available technology.

One cannot help but notice that Soviet theoretical sciences such as mathematics and physics stood at world level but actual engineering lagged somewhat behind latest Western developments. Having said that, even today attaining some of the goals the Soviets aimed for would be technically difficult and very costly. Thus, while the Soviets felt the need for speed and endurance, the need could not be satisfied for a number of reasons, foremost technical ones.

The same M-50 prototype with a different 'bort', 023 (see also the colour section). As spectacular as this big jet was, it was a failure and never attained the desired performance.

A Tupolev Tu-22 supersonic bomber, with its swept wings, conveyed an impression of speed and dynamic, through in fact – and as admitted by its designer – it was one of his less-fortunate creations.

While advanced versions of the Tu-22 went on to serve as a missile carrier for the DA, earlier Tu-22s and Tu-22Rs retained nuclear strike with bombs as their primary role.

models were the Tu-22B (designed for delivery of free fall bombs), the Tu-22K (missile carrier), and the Tu-22R (reconnaissance variant).

While not the subject matter of this work, it is nevertheless appropriate to mention the *Aviatsiya Voyenno-Morskogo Flota* (AV-MF) (Soviet Naval Aviation). Since maritime reconnaissance and strike, as well as patrol and anti-submarine missions, require long-range and large payloads, the AV-MF shared many aircraft types with the Long-Range Aviation, such as the Tu-16. The aircraft and the weapons they carried were optimised for maritime operations and naval attack yet they shared much, as was illustrated in the sub-chapter dealing with the test of the nuclear-tipped potentially carrier-killing anti-ship missile.

The Achilles Heel

Desite the testing of weapons and the development of aircraft, the *Dalnaya Aviatsiya* continued suffering from a number of weaknesses. Arguably the greatest was its limited range, which excluded anything but small-scale attacks against the US mainland. There were two reasons for this: technical limitations, foremost the limited range of the aircraft in service steaming from engine performance, fuel consumption versus fuel capacity and similar factors; and the geographic and political limitations, as airfields in the Soviet Union were too far away from potential targets in the US. No matter how well developed Soviet nuclear weapons were with aircraft capable of their delivery and airmen proficient in utilising them – all this was of little use if the principal adversary's territory could not be reached. Yet despite their clear disadvantages the Soviets were not contemplating this in resignation, but instead various solutions were put forward and tried out.

One solution was to develop an in-flight refuelling capability, which the Soviets did working out a number of methods of in-flight fuel transfer and modified several types of bombers for that purpose. The long-range capability thus acquired was tested in a number of exercises. For example, a 3M bomber construction number 0204 flew 14,660km, being twice refuelled in the air from a M-4 tanker (construction number 2528). The mission described was a 'round trip' with take off and landing at Ramenskoye, flown 20-21 May 1958 – since the flight lasted for 18 hours and 8 minutes it started on one day and finished the next.[7] The 3M thus became the first Soviet bomber with true intercontinental range.

An even longer mission was flown a year later in the spring of 1959 by a pair of Tu-95 bombers of the 1006 TBAP which took off from Uzin in the Ukraine, dropped mock 'nuclear bombs' in the Arctic and flew back to base. Available data differs somewhat on the exact distance and duration of the flight – 16,950km and 17,150km being given – with the time in any case exceeding 21 hours. The mission was personally led by the CO of the 106th TBAD, General V. V. Reshetnikov, who piloted one of the aircraft.[8] Myasishchev bombers again demonstrated their long-range when on 1 October 1965 three 3M bombers from Ukrainka flew up to St. Matthew Island and returned to base. This mission lasted 11 hours 30 minutes and was accomplished with one in-flight refuelling.[9] In addition to in-flight refuelling range could also be extended by utilising outlying airfields such as Olenogorsk, Rogachevo, Tiksi and Ugolny for staging, refuelling and servicing. These were referred to as *aerodromy podskoka* or hoping airfields for they facilitated 'hoping' closer to an intended target.

The development of in-flight refuelling capability, was certainly useful, as was also the use of 'hoping airfields' but all this alleviated the range problem to only a limited extend. What was really needed were bases much closer to potential targets in the United States. This was obvious but the solution was not if no one was willing to host them. The answer was that if no bases on land were available ones on water might provide the solution. The water would be frozen though. Large ice fields in the Arctic were natural aerodromes: firm, flat and long surfaces mostly devoid of any obstacles. Aircraft supporting Soviet Arctic exploration had landed and taken off from the ice, furnishing proof of the practical feasibility of the idea. It appears that the first instance of Long-Range Aviation bombers landing and taking off from Arctic ice took place as part of Operation Sever-5 (North-5) conducted in April-May 1950. In course of the undertaking, two Tu-4 bombers with pilots Vagapov and Simonov at the controls landed and took off from an ice field with the second aircraft acting as a refueller. More modern heavy bombers attempted the same several years later when on 23 May 1958 a Tu-16 with Colonel A. Alekhnovich at the controls and a Tu-95 piloted by Major N. Bazarnog landed on Arctic ice. All seemed to indicate that long-range bomber operations could be conducted in this fashion until a mishap proved otherwise. When taking off, the Tu-16 collided with an Il-14 transport suffering irreparable damage. The whole affair was rather embarrassing, especially given that the wrecked Soviet bomber was spotted by a Canadian reconnaissance aircraft. Thus, the only thing left to be done was to strip the stricken aircraft of engines and any other usable parts and subsequently torch what remained. With this, Soviet hopes of strategic bombers operating from the Arctic ice literally went up in smoke. In due time as the drifting ice field melted the charred wreckage sank to the bottom of the ocean. As far as it is known the Soviets did not attempt any more such operations and the only 'red heavies' operating from Arctic ice would be large transports supporting Soviet civilian scientific and exploratory missions in those parts of the world.[10]

Summarising the above in a succinct manner it seems fitting to say that while the Soviets developed an in-flight refuelling capability and tried various operational concepts – such as the hoping airfields and Arctic ice operations – it still eluded them to attain a truly intercontinental reach enabling their bombers to strike the US mainland at will, and this remained the Long-Range Aviation's Achilles heel.

Switch to Missiles and Low Altitude

Ironically, at about the time that the DA reached its peak power, in 1961-1962, all the work on new Soviet intercontinental bombers was brought to a virtual halt, and design bureaus were ordered to concentrate on improving the available missile systems, or creating new missiles for existing aircraft. Never in a position to afford weapon systems that were unlikely to prove effective in combat, and concluding that the traditional high-altitude operations by strategic bombers were unlikely to be successful in the face of air defences now protecting not only the USA, but the whole of Canadian airspace too, the government of Soviet leader Nikita Khrushchev decided to replace bombers with ballistic missiles as the primary means of deploying nuclear weapons. In turn, in 1964, changes in the political leadership of the USSR led to another complete revision of the basic premise of military doctrine: the new government arrived at the decision to achieve quantitative and qualitative parity with the USA – essentially the decision that ultimately ruined the Soviet Empire. Orders were orders though, and thus by 1967 the DA requested a new intercontinental bomber equipped with long-range Kh-45 cruise missiles, and short-range anti-air defence missiles. However, not only did the Long-Range Aviation then object to the involvement of the Sukhoi Design Bureau – demanded by the highest Soviet military authority and sole authority responsible for armament acquisition, the General Staff of the Soviet Armed Forces (GenStab) – but the

Although remaining practically unable to reach their primary targets in the USA – except on one-way missions – Soviet heavy bombers, like this Tu-95, nicely illustrating its huge (and extremely loud!) contra-rotating propellers, continued serving through the 1960s and then into the 1970s.

related research and development was delayed and then, in 1970, the requirement was completely re-written. Ultimately, the contest between the Myasischev M-18, Sukhoi T-4MS and the Tupolev Designs 132 and 135 was decided in favour of the latter, though modified through the addition of a variable-geometry wing. This led to permission for development of what became the Tu-160, the last Soviet intercontinental bomber, granted by Moscow only in 1975. It is sufficient to say that the Tu-160 made its first flight in December 1981, but entered service only in May 1987, and was then equipped with Kh-55 cruise missiles instead of the Kh-45s.

Meanwhile, in response to the block-obsolescence of aircraft and armament, the GenStab ordered the Long-Range Aviation to start working on tactics that could be used to penetrate enemy air defences. Correspondingly, through the early 1960s the DA began experimenting with low altitude operations of 3M and Tu-95 bombers. Moreover, an entirely new family of air-launched cruise missiles entered research and development – almost exclusively as weapons for reconfigured Tu-16s, Tu-22s and Tu-95 bombers. However, the times of not only frequent, but almost daily live testing of Soviet nuclear weapons – indeed: the times of atmospheric nuclear tests – was over by then: thus, contrary to the earlier KS-1s, K-10s, KSR-2s and KSR-5, none of the new, supersonic missiles of the Kh-22 series was ever tested while equipped with a nuclear warhead.

AFTERWORD – WE ARE STILL HERE

We think of ourselves as living in times of rapid progress. However, the 1940s-1960s period also saw such a period. In 1951, the Soviets for the first time dropped a nuclear bomb from a piston engined bomber, meaning they were – essentially – at the level attained by the United States at the end of Second World War. Yet in little over a decade they had literally hundreds of jet and turboprop bombers capable of delivering nuclear and thermonuclear bombs as well as cruise missiles with nuclear warheads. In addition, they also developed an in-flight refuelling capability. This cannot be seen as anything but very rapid progress in a short period of time. Certainly, the Soviet bomber force organised into the Long-Range Aviation was for all practical purposes incapable of anything more than token, attacks against targets in the continental United States, and then probably one-way only. While this would obviously not suffice to 'win' a nuclear war or assure mutual destruction, even one or two nuclear or thermonuclear bombs could produce casualties in the range of hundreds of thousands if not millions. This would not come even close to threatening America's survival but would be very painful in particular to those directly affected. That said, the situation for Western Europe and US allies in the Middle East and Far East could look much different, as many targets, among them large population centres, would be within the Soviet nuclear bomber force's striking range, in particular when in-flight refuelling is accounted for. Certainly, retaliatory nuclear strikes could effectively wipe out the Soviets and their satellites states: that, however, would be of little comfort to the dead and maimed, and it could neither restore nor decontaminate what was destroyed and irradiated. Suffice to say while North America would in all likelihood escape with limited damage (keeping in mind that the timeframe is narrowed down to the period before the Soviet ICBM force was built up) much of both of the rest of the Western and Eastern world would be reduced to nuclear ashes. The world was, of course, spared of such fate in the 1960s and later on, and we are still here.

Deaf and Dumb

Test dropping live nuclear bombs was not a routine peacetime military task and as such was not – bar some exceptions – carried out by regular aviation units. Instead the *71-y Spetsialny Poligon* (71st Special Polygon) known also as the *Voyskovaya Chast 93851* (Military Unit 93851), was entrusted with these missions. In Eastern Bloc military terminology a 'polygon' usually meant an exercise or manoeuvre area, a trials field or some such. In this case however it was used to designate a special aviation unit, the main task of which was to live test nuclear aerial bombs.

The 71st Special Polygon was based at Bagerovo airfield located some 14km northwest of Kerch City on the Kerch Peninsula, itself part of the Crimean Peninsula. It was composed of three sub-units: the *35-y Otdelny Spetsialny Ispytatelny Aviatsionny Polk* (35th Independent Special Test Aviation Regiment), the *513-y Istrebitelny Aaviatsionny Polk* (513th Fighter Aviation Regiment), and, finally, the *647-y Smeshannyy Aviatsionnyy Polk Spetsialnogo Obespecheniya* (647th Mixed Special Support Aviation Regiment).[11] The 35th Regiment was in fact a special bomber unit which had in its inventory all the main Soviet bomber types such as the Tu-4, Tu-16, Tu-95 and Il-28, which were configured for nuclear weapon delivery. Its mission was to conduct live test drops of nuclear bombs. In turn the 513th Regiment which initially flew La-11s, but subsequently re-equipped with MiG jets, was tasked with providing fighter escort for the bombers and airspace security during the tests. Last but not least, the 647th Regiment which had on charge a variety of propeller driven and jet powered fixed-wing aircraft, and some helicopters, was entrusted with a large variety of auxiliary yet vital tasks such as collecting air samples for radiological analysis, filming and photographing the tests, as well as the more mundane jobs of ferrying personnel and material.

Aircraft and men of the 71st Special Polygon conducted 178 live tests of nuclear and thermonuclear weapons: 94 at Semipalatinsk, 83 at Novaya Zemlya and a single one at Totskoye. Yet military secrecy required the personnel of the unit to pretend not to hear or know anything. One might add being mute as well, for very little was known about all this activity until the demise of the Soviet Union and the end of the Cold War. As shown in Table 5, the intensiveness of the testing in question often reached what are nowadays entirely unimaginable scales: in several periods during the late 1950s, and again in 1962 in particular, the crews of the 71st Special Polygon were test dropping up to three or four nuclear bombs a day!

Pre-flight briefing of a Tu-16-crew of the 35th Independent Test Aviation Regiment. (Albert Grandolini Collection)

Even years after the end of the live nuclear testing of the 1950s and the 1960s, for most of their operational service, Tu-4s, Tu-16s, and Tu-95s of the 35th Independence Test Aviation Regiment, 71st Special Polygon, retained their anonymous appearance: even 'bort' numbers were rarely applied on them. It was only during the late 1980s that some of the aircraft began wearing 'special' insignia. (Albert Grandolini Collection)

Table 4: DA Test-Units involved in Soviet Nuclear Testing, 1949-1963

Unit and Sub-Units	Base	Equipment	Notes
71st Special Polygon	Bagerovo AB (Kerch, Crimean Peninsula)		also known as 'Military unit 93851'
35th Independent Test Aviation Regiment		Tu-4, Tu-16, Tu-95, Il-28	bomber unit, all aircraft configured for deployment of nuclear weapons
513th Fighter Aviation Regiment		MiG-15bis, MiG-17F, MiG-19S, Su-9	tasked with providing top cover for bombers involved in nuclear tests, and airspace security during testing
647th Special Support Aviation Regiment		Li-2, Il-14, Tu-4, Tu-16, Tu-95	collecting samples for radiological analysis, filming and photographing tests, ferrying personnel and equipment

Table 5: Air dropped Soviet Nuclear Tests, 1949-1962[12]

Date	US-Code	Weapon	Notes
29 Aug 1949	Joe 1	RDS-1	Semipalatinsk, tower, surface detonation
18 Oct 1951	Joe 3	RDS-3	Semipalatinsk; first bomb drop test from Tu-4
12 Aug 1953	Joe 4	RDS-6	Semipalatinsk, tower, surface detonation; first thermonuclear weapon
23 Aug 1953	Joe 5	RDS-4	Semipalatinsk, first production tactical nuclear bomb; air drop from Tu-4
14 Sep 1954	Joe 8	RDS-3	Orenburg, Exercise Snezhok with real nuclear detonation, air drop from Tu-4
30 Oct 1954	Joe 14		Semipalatinsk, air drop from Tu-4
6 Nov 1955	Joe 18	RDS-27	Semipalatinsk, air drop
22 Nov 1955	Joe 19	RDS-37	Semipalatinsk, air drop; first hydrogen bomb; air drop from Tu-16; 2 KIA, 47 WIA
30 Aug 1956	Joe 24	RDS-37	Semipalatinsk, air drop from Tu-16
2 Sep 1956	Joe 25		Semipalatinsk, air drop from Tu-16
10 Sep 1956	Joe 26		Semipalatinsk, air drop from Tu-16; live exercise with 27 Mi-4 helicopters and 272 troops
17 Nov 1956	Joe 27		Semipalatinsk, air drop
14 Dec 1956	Joe 28		Semipalatinsk, air drop
8 Mar 1957	Joe 30		Semipalatinsk, air drop
3 Apr 1957	Joe 31		Semipalatinsk, air drop
6 Apr 1957	Joe 32		Semipalatinsk, air drop
10 Apr 1957	Joe 33		Semipalatinsk, air drop
12 Apr 1957	Joe 34		Semipalatinsk, air drop
16 April 1957	Joe 35	TN	Semipalatinsk, air drop
22 Aug 1957	Joe 36	TN	Semipalatinsk, air drop
26 Aug 1957			Semipalatinsk, air drop
24 Sep 1957	Joe 39		Semipalatinsk, air drop
26 Sep 1957	Joe 40	TN	Semipalatinsk, air drop
6 Oct 1957	Joe 41	RDS-46A	Semipalatinsk, air drop from Tu-16
28 Dec 1957	Joe 43		Semipalatinsk, air drop
4 Jan 1958	Joe 44		Semipalatinsk, air drop
17 Jan 1958	Joe 45		Semipalatinsk, air drop
23 Feb 1958	Joe 46	TN	Novaya Zemlya, air drop
27 Feb 1958	Joe 47		Semipalatinsk, air drop
27 Feb 1958	Joe 48		Novaya Zemlya, air drop
13 Mar 1958	Joe 49		Semipalatinsk, air drop
13 Mar 1958	-		Novaya Zemlya, atmospheric, unclear if air drop or rocket/missile
14 Mar 1958	Joe 51		Semipalatinsk, air drop
14 Mar 1958	Joe 50		Semipalatinsk, air drop

Table 5 *(continued)*			
Table 5: Air dropped Soviet Nuclear Tests, 1949-1962[12]			
15 Mar 1958	Joe 52		Semipalatinsk, air drop
15 Mar 1958	-		Semipalatinsk, atmospheric, unclear if air drop or rocket/missile
18 Mar 1958	-		Semipalatinsk, air drop
20 Mar 1958	Joe 53		Semipalatinsk, air drop
21 Mar 1958	Joe 54		Novaya Zemlya, air drop
22 Mar 1958	Joe 55		Semipalatinsk, air drop
30 Sep 1958	Joe 56		Novaya Zemlya, air drop
2 Oct 1958	Joe 57		Novaya Zemlya, air drop
2 Oct 1958	Joe 58		Novaya Zemlya, atmospheric, unclear if air drop or rocket/missile
5 Oct 1958	Joe 61		Novaya Zemlya, air drop
6 Oct 1958	Joe 62		Novaya Zemlya, air drop
10 Oct 1958	Joe 63	TN	Novaya Zemlya, air drop
12 Oct 1958	Joe 64	TN	Novaya Zemlya, air drop
15 Oct 1958	Joe 65	TN	Novaya Zemlya, air drop
18 Oct 1958	Joe 66	TN	Novaya Zemlya, air drop
19 Oct 1958	Joe 67		Novaya Zemlya, air drop
19 Oct 1958			Novaya Zemlya, air drop
20 Oct 1958	Joe 68	TN	Novaya Zemlya, air drop
21 Oct 1958	Joe 69		Novaya Zemlya, air drop
22 Oct 1958	Joe 70	TN	Novaya Zemlya, air drop
24 Oct 1958	Joe 71	TN	Novaya Zemlya, air drop
25 Oct 1958	-	TN	Novaya Zemlya, air drop
1 Sep 1961	Joe 75		Semipalatinsk, air drop
4 Sep 1961	Joe 76		Semipalatinsk, air drop
5 Sep 1961	Joe 77		Semipalatinsk, air drop
6 Sep 1961	Joe 78		Semipalatinsk, air drop
10 Sep 1961	Joe 80		Novaya Zemlya, air drop from Tu-95 on 'Battlefield D-2' in Mityushika Bay
13 Sep 1961	Joe 84		Semipalatinsk, air drop
17 Sep 1961	Joe 88		Semipalatinsk, air drop
18 Sep 1961	-		Semipalatinsk, atmospheric
18 Sep 1961	Joe 89		Novaya Zemlya, unclear if air drop or missile test
20 Sep 1961	-		Semipalatinsk, air drop
21 Sep 1961	Joe 92		Semipalatinsk, air drop
26 Sep 1961	-		Semipalatinsk, air drop
1 Oct 1961	-		Semipalatinsk, air drop
2 Oct 1961	Joe 94		Novaya Zemlya, air drop
4 Oct 1961	Joe 95		Semipalatinsk, air drop
6 Oct 1961	Joe 97		Novaya Zemlya, air drop
8 Oct 1961	Joe 99	K-10S	Novaya Zemlya; cruise missile, test of the K-10 warhead from Tu-16K
12 Oct 1961	Joe 101		Semipalatinsk, air drop
17 Oct 1961	Joe 102		Semipalatinsk, air drop
19 Oct 1961	Joe 103		Semipalatinsk, air drop
23 Oct 1961	Joe 106		Novaya Zemlya, air drop
25 Oct 1961			Semipalatinsk, air drop
25 Oct 1961	Joe 108	TN	Novaya Zemlya, air drop

Table 5 *(continued)*			
Table 5: Air dropped Soviet Nuclear Tests, 1949-1962[12]			
30 Oct 1961	Joe 111	RDS-202	Novaya Zemlya, air drop from Tu-95V; 'Tsar Bomba', largest man-made detonation ever
31 Oct 1961	Joe 112		Novaya Zemlya, air drop
31 Oct 1961	Joe 113		Novaya Zemlya, air drop
1 Nov 1961	Joe 114		Semipalatinsk, air drop
2 Nov 1961	Joe 116	TN	Novaya Zemlya, air drop
2 Nov 1961	-		Semipalatinsk, air drop
3 Nov 1961	Joe 117		Semipalatinsk, air drop
4 Nov 1961	Joe 118		Novaya Zemlya, air drop
4 Nov 1961	-		Novaya Zemlya, air drop
4 Nov 1961	Joe 119		Novaya Zemlya, air drop
27 Feb 1962			Novaya Zemlya, air drop from M3
5 Aug 1962			Novaya Zemlya, air drop from Tu-95V
10 Aug 1962			Novaya Zemlya, air drop
18 Aug 1962			Semipalatinsk, air drop; 150th Soviet nuclear weapon test
20 Aug 1962			Novaya Zemlya, unclear if air drop or missile test
21 Aug 1962			Semipalatinsk, air drop
22 Aug 1962			Semipalatinsk, air drop
22 Aug 1962		K-10S	Kola Peninsula; cruise missile from Tu-16K; alternative reports indicate KSR-2
22 Aug 1962			Novaya Zemlya, air drop
25 Aug 1962			Semipalatinsk, air drop
25 Aug 1962			Novaya Zemlya, air drop
27 Aug 1962			Semipalatinsk, toss from Su-7B fighter-bomber
27 Aug 1962			Novaya Zemlya, air drop
2 Sep 1962			Novaya Zemlya, air drop
15 Sep 1962			Novaya Zemlya, air drop from Tu-16A; exercises by 132 TBAP and 185 TBAP
16 Sep 1962			Novaya Zemlya, air drop from Tu-16A; exercises by 132 TBAP and 185 TBAP
18 Sep 1962			Semipalatinsk, air drop
18 Sep 1962			Novaya Zemlya, air drop
19 Sep 1962			Novaya Zemlya, air drop
21 Sep 1962			Novaya Zemlya, air drop
24 Sep 1962			Semipalatinsk, air drop

Table 5 (*continued*)			
Table 5: Air dropped Soviet Nuclear Tests, 1949-1962[12]			
25 Sep 1962			Novaya Zemlya, air drop
27 Sep 1962			Novaya Zemlya, air drop
28 Sep 1962			Semipalatinsk, air drop
7 Oct 1962			Novaya Zemlya, air drop
9 Oct 1962			Semipalatinsk, air drop
9 Oct 1962			Novaya Zemlya, air drop
13 Oct 1962			Semipalatinsk, air drop
14 Oct 1962			Semipalatinsk, air drop
20 Oct 1962			Semipalatinsk, air drop
22 Oct 1962			Novaya Zemlya, air drop
27 Oct 1962			Novaya Zemlya, air drop
28 Oct 1962			Semipalatinsk, air drop
30 Oct 1962			Novaya Zemlya, air drop
31 Oct 1962			Semipalatinsk, air drop
1 Nov 1962			Novaya Zemlya, air drop
1 Nov 1962			Semipalatinsk, air drop
3 Nov 1962			Novaya Zemlya, air drop
3 Nov 1962			Semipalatinsk, air drop
14 Nov 1962			Semipalatinsk, air drop
17 Nov 1962			Semipalatinsk, air drop
18 Dec 1962			Novaya Zemlya, air drop
20 Dec 1962			Novaya Zemlya, air drop
22 Dec 1962			Novaya Zemlya, air drop
23 Dec 1962			Novaya Zemlya, air drop
23 Dec 1962			Novaya Zemlya, air drop
23 Dec 1962			Novaya Zemlya, air drop
24 Dec 1962			Novaya Zemlya, air drop from Tu-95V
24 Dec 1962			Novaya Zemlya, air drop
25 Dec 1962			Novaya Zemlya, air drop
25 Dec 1962			Novaya Zemlya, air drop; last Soviet atmospheric test (No. 222 in total)

Bunker-42: Doomsday Dungeon

Apart from aircraft and airfields, the Long-Range Aviation also needed other facilities. Before getting to the point it is worth mentioning the Moscow metro or subway which has the primary peacetime function of facilitating the daily movement of the city's large population. However it had since the time of its construction also a secondary wartime military role. During the Great Patriotic War as in the Soviet Union and nowadays Russia the war on the Eastern Front was called the metro served as an air raid shelter for the general population but also as an underground command post or rather posts. Namely the Kirovskaya (currently Chistye Prudy) station was taken over by the General Staff while commands of lesser importance were located at Krasnye Vorota, Belorusskaya and Ploshchad Sverdlova (now Teatralnaya) stations.

With the end of the Second World War universal peace did not come about and the world found itself embroiled in a new conflict: the Cold War. Obviously, this was different in nature to the preceding war and Muscovites did not fear on a daily basis that air raid sirens will compel them to seek shelter in the metro. Nevertheless, it was clear that the situation could escalate from 'cold' into 'hot' both figuratively as well as literally. For this reason, it should not come as a surprise that as the Moscow subway system was expanded, additional dual-use or strictly military underground facilities were also built.

One such facility had its entrance in the vicinity of Taganskaya subway station.[6] Even before this station was built in 1950, it was decided to locate a military underground facility in its immediate proximity. Accordingly, four main tunnels nine metres wide and 100 metres long were dug at a depth of 65 metres, as well as auxiliary passages, shafts and other installations. Work was completed in 1956 with the facility receiving a numeric designation 42 which meant little to any outsider, this being purposely so. This was, in fact, a bunker fitted for one-month-long autonomous operations and until 1986 housed the Long-Range Aviation's wartime command post. A few years after its completion, 'Bunker-42' came very close to being used for its intended purpose. In 1962, during the Cuban Missile Crisis, it was fully staffed by 600 officers and men who were ready to command Soviet bombers but fortunately it did not come to that.

Once the Soviet military vacated the bunker it was utilised as a civilian communications hub but operating the site was neither cost effective nor very practical and it was finally closed down in 2006. Since then, it has become a museum open to the public and anyone curious enough can nowadays access this Soviet doomsday dungeon.

The entrance to the Bunker-42 – the Long-Range Aviation's wartime command post, in the vicinity of Taganskaya subway station in Moscow. Having lost its military role, it now serves as a museum open to the public. (Courtesy: Maria Ulanova, Bunker-42 Cold War Museum)

The underground corridor at the entrance to the Bunker-42, the 'Doomsday Dungeon'. In 'old times' it would have been anything than easy to pass the officer seated at that desk and to enter one of four major halls of the complex. Notably, the furniture and carpets have a typically Soviet aesthetics. (Courtesy: Maria Ulanova, Bunker-42 Cold War Museum)

BIBLIOGRAPHY

Alekhin, R., *Vozdushno-desantnyye voyska: istoriya rossiyskogo desanta* (EKSMO, 2009)

Andryushin, I. A., Chernyshev, A. K., Yudin, Yu. A., *Ukroshcheniye yadra* (Sarov, 2003)

Grimes, B., *The History of Big Safari* (Archway Publishing, 2014)

Golubev, Yu. V., *Na ostriye Atomnogo proyekta* (Chelyabinsk, 2015)

Kulikov, S. M., *Istoriya VNIIA v litsa* (Moskva: TSNIIatominform, 2014)

Kulikov, S. M., *Aviatsiya i yadernyye ispytaniya* (Moskva: TSNIIatominform, 1998)

Markovskiy, V., Perov, K., *Sovetskiye aviatsionnyye rakety «vozdukh-zemlya»* (Moskva: Eksprint)

Mikhailov, V. N., *Yadernyye ispytaniya SSSR* (Sarov, 1997)

Mikhailov, V. N., *Yadernyye ispytaniya t. 1: Arkticheskiy yadernyy polygon* (Moskva: TSNIIatominform, 2006)

Mikhailov, V. N., *Catalog of Worldwide Nuclear Testing* (Begell House Inc., 1999)

Yakubovich, N. V. *Myasishchev. Neudobnyy geniy. Zabytyye pobedy sovetskoy aviatsii* (Eksmo, 2008)

Zelentsov, S. A. *Yadernyye ispytaniya. Totskoye voyskovoye ucheniye (nauchno-publitsisticheskaya monografiya)* (Moskva, 2006)

West, N. & Tsarev, O., *The Crown Jewels: The British Secrets at the Heart of the KGB Archives* (Yale University Press, 1999)

Magazine and Internet Articles

Botev, V. M., 'Ministerstvo oborony v Atomnom proyekte', *Atom*, 2019-2.

Medved, A. N. K., 'Istorii sozdanya pervoy otechestvennoy yadernoy bomby' www.engine.aviaport.ru

Moroz, S. 'Bombardirovshchik Tu-95' (www.naukatechnika.com).

Nikolayev, A., 'Dalnaya Aviatsiya Samolety i Lyudi' (www.engine.aviaport.ru).

Oleynikov, P. V., 'German scientists in the Soviet atomic project', *Nonproliferation Review*, Volume 7, No 2, Summer 2000.

Saperov, V., Roshchin, G., 'V zaoblachnyye predely... Vorvalos' yarchaysheye svecheniye: Epizod poluvekovoy davnosti iz epokhi kholodnoy voyny Voyenno-promyshlennyy', *Kuryer*, No 23 (440) 2012.

Central Intelligence Agency, National Intelligence Estimate 11-2A-62, "Soviet Atomic Energy Program", 16 May 1962, (www.cia.gov).

NSA, official Venona site, (www.nsa.gov).

NOTES

Chapter 1

1 A. N. K. Medved, 'Istorii sozdanya pervoy otechestvennoy yadernoy bomby'
2 N. West & O. Tsarev, The Crown Jewels: The British Secrets at the Heart of the KGB Archives.
3 Andryshin, Ukroshcheniye yadra
4 Nonproliferation Review Volume 7, No 2, Summer 2000
5 https://www.nsa.gov/news-features/declassified-documents/venona/
6 Andryshin.
7 The ASCC was a commission consisting of experts from air forces of the 'Five Eyes' countries (Australia, Canada, New Zealand, the United Kingdom and United States), responsible for furthering each country's capabilities for combined cooperation and for assigning names to foreign weapon systems. The same system was adopted by the North Atlantic Treaty Organisation (NATO).
8 Unless stated otherwise, this chapter is based upon S.M. Kulikov *Aviatsiya i yadernyye ispytaniya.*
9 www.warherous.ru

Chapter 2

1 S.A. Zelentsov, *Yadernyye ispytaniya. Totskoye voyskovoye ucheniye (nauchno-publitsisticheskaya monografiya)*
2 Unless stated otherwise, this chapter is based upon R. Alekhin, *Vozdushno-desantnyye voyska: istoriya rossiyskogo desanta.*

Chapter 3

1 Unless stated otherwise, this chapter is based upon Mikhailov, V. N., *Yadernyye ispytaniya SSSR*, 1997.
2 *Physics Today* Volume 70, Issue 1, January 2017.
3 www.warherous.ru
4 Dreyfuss, *Symbol Sourcebook: An Authoritative Guide to International Graphic Symbols*

Chapter 4

1 This was a reference to the exclamation by Nikita Khrushchev '*My vam pokazhem Kuzkinu mat!*', which literally means 'we will show you Kuzma's mother', but colloquially expresses the intention to teach others a harsh lesson.
2 Unless stated otherwise, this chapter is based upon S.M. Kulikov, *Istoriya VNIIA v litsa*
3 Kulikov, *Aviatsiya i yadernyye ispytaniya*
4 B. Grimes, *The History of Big Safari.*
5 Yu. V. Golubev, *Na ostriye Atomnogo proyekta*
6 The Tu-144 was colloquially referred to as the 'Konkordski' due to its external similarity to the Anglo-French Concorde, even though the design of the two aircraft differed considerably.
7 www.warherous.ru

Chapter 5

1 N.V. Yakubovich, *Myasishchev. Neudobnyy geniy. Zabytyye pobedy sovetskoy aviatsii*
2 V. Markovskiy, Perov K. *Sovetskiye aviatsionnyye rakety «vozdukh-zemlya»*
3 M. Pervov, *Otechestvennoye raketnoye oruzhiye*
4 Central Intelligence Agency, National Intelligence Estimate 11-2A-62
5 V.N. Mikhailov, *Catalog of Worldwide Nuclear Testing*
6 *Atom* 2019-2. At this point the author cannot but ponder the fact that information on certain events in available Russian sources is in some cases profuse with accurate details while in others very scant and conflicting. It is most difficult to make out a pattern as this equally applies to setbacks as well as successes, while the events in question have – due to the passage of time – absolutely no intelligence or other practical relevance except for historians, aviation enthusiasts and the like to require that they keep them under a mantle of secrecy.
7 Mikhailov, *Catalog of worldwide nuclear testing*
8 V.N. Mikhailov, *Yadernyye ispytaniya t. 1 : Arkticheskiy yadernyy polygon*
9 www.airwar.ru
10 *Tayfun No 12/2012*
11 As a side note it is worth adding that the Soviets also live tested other naval weapons with nuclear warheads, such as T-5 submarine launched torpedoes with RDS-9 warhead: that, however, is a story which deserves to be described in detail separately.
12 *Voyenno-promyshlennyy kuryer No 23* (440) 2012
13 The word 'tactical' in this context meaning that they were to practice the tactics of thermonuclear bomb delivery.
14 Voyenno-promyshlennyy kuryer No 44 (807) 2019
15 Notably, because the letter 'z' in the Cyrillic alphabet is written 'з' the aircraft has been erroneously referred to as the Tu-163 in some Western publications.

Chapter 6

1 Unless stated otherwise the source for the orders of battle given is *www.engine.aviaport.ru*
2 *www.da-sv.ru*
3 *www.naukatechnika.com*
4 Yakubovich, *Myasishchev.*
5 Note: the 1225 and 1229 TBAPs were at that time directly under the control of the corps commanders but would in due time form the core of the new 31 TBAD
6 http://bunker42.com
7 Yakubovich, *Myasishchev.*
8 www.naukatechnika.com
9 Yakubovich, *Myasishchev.*
10 *Aviatsija i* Vremya 2001/01
11 The Russian word *obespecheniye* literally means 'security' or 'securing' but, in this context, particularly considering the unit's mission, to translate it as support seems most fitting.
12 Entries in bold are confirmed air drops; TN stands for thermonuclear.

ACKNOWLEDGEMENTS

The author would like to express his most sincere gratitude to Tom Cooper and Dmitry Zubkov – without their encouragement as well as practical help the present volume would have never come into being.

ABOUT THE AUTHOR

Krzysztof Dabrowski from Poland has a lifelong interest in the subject of military aviation, and has written dozens of articles on a variety of related subjects for printed magazines and the ACIG.org/ACIG.info, AeroHisto and The Boresight websites. His particular area of interest is air warfare during the Cold War, the aircraft involved and experiences of their crews